Infinite Minds

Infinite
Minds
A Philosophical
Cosmology

John Leslie

CLARENDON PRESS · OXFORD

OXFORD
UNIVERSITY PRESS

Great Clarendon Street, Oxford OX2 6DP

Oxford University Press is a department of the University of Oxford.
It furthers the University's objective of excellence in research, scholarship,
and education by publishing worldwide in

Oxford New York

Athens Auckland Bangkok Bogotá Buenos Aires Cape Town
Chennai Dar es Salaam Delhi Florence Hong Kong Istanbul Karachi
Kolkata Kuala Lumpur Madrid Melbourne Mexico City Mumbai Nairobi
Paris São Paulo Shanghai Singapore Taipei Tokyo Toronto Warsaw
with associated companies in Berlin Ibadan

Oxford is a registered trade mark of Oxford University Press
in the UK and in certain other countries

Published in the United States
by Oxford University Press Inc., New York

British Library Cataloguing in Publication Data
Data available

Library of Congress Cataloging in Publication Data
Leslie, John, 1940–
Infinite minds: a philosophical cosmology/John Leslie.
p. cm.
Includes bibliographical references (p.) and indexes.
1. Cosmology—Philosophy. I.. Title.
BD511.L475 2001   113—dc21—2001036482
ISBN 0-19-924892-3

10 9 8 7 6 5 4 3 2 1

Typeset in Calisto MT by
Cambrian Typesetters, Frimley, Surrey

Printed in Great Britain
on acid-free paper by
T.J. International Ltd
Padstow, Cornwall

Preface

Infinite Minds returns to various themes introduced in *Value and Existence* (1979), a book achieving minor notoriety when the influential philosopher J. L. Mackie wrote a chapter of his *The Miracle of Theism* around it, saying it had outlined 'a formidable rival' to traditional theism. 'If, with Leibniz and others, you demand an ultimate explanation, then this may well be a better one', he concluded. The book's central idea was that an abstract factor, The Good of Plato, might itself bear creative responsibility for the cosmos. Present in Plato's *Republic*, it was an idea in which Mackie could find no actual contradiction. He may have been wrong in thinking that traditional theists had never defended anything on these lines, and he argued in any case that the idea fared badly because good and bad were simply human fabrications, as explained in his *Ethics: Inventing Right and Wrong*. Still, I remain extremely grateful to him for drawing attention to my work.

Apart from very greatly expanding various earlier points and mentioning others only briefly, I now have a major change of emphasis. *Value and Existence* maintained Neoplatonically that 'God' might be a name for a creative force, a force of *creative ethical requirement* or—just another way of saying the same thing—a principle that consistent groups of ethical requirements, ethical demands for the actual presence of this or that situation, can sometimes bring about their own fulfilment. The cosmos might exist because its existence was ethically necessary, without the aid of an omnipotent being who chose to do something about this. Now, although I also wrote that a divine person might well head the list of the things that the creative force would have created, readers often overlooked this, or they treated it as an unfortunate concession to please religious folk. Again, they tended not to realize that my preferred picture, reached towards the end of the book's

Chapter 11, of a cosmos consisting of infinitely many unified realms of consciousness, each of them infinitely rich, might be described as *a pantheistic picture* of an unusual type: a picture *of infinitely many minds each worth calling 'divine'*. All this was largely my fault and the pages now in your hands try to remedy matters. 'An Infinite Mind' would have been quite the wrong title for them. However, they allow themselves to speak of 'the' divine mind, much as believers in many actual universes often talk of 'the' universe. After all, islanders who believe in many islands can still refer to 'the island', meaning their very own.

Other new things include these:

1. I now emphasize that each infinitely rich realm of consciousness (or 'divine mind' if you agree to call it that) could be expected to include knowledge of absolutely everything worth knowing. *Exactly how it feels to be people such as you and me* might well be one thing worth knowing, so I see no huge problem in reconciling my pantheism with the plain facts of experience. There could be the following difficulty, though. Very disorderly experiences might be considered worth knowing. In that case, shouldn't one expect one's own experiences to become disorderly at any moment? This can seem rather a serious difficulty but I try to defeat it.

2. Earlier I wrote that, since only consciousness has intrinsic worth, 'intrinsic ethical requiredness', it might well follow that sticks, stones and stars of the kinds in which people ordinarily believe *would be in some sense fictions*, although ones of a very useful sort. I now suggest instead that a divine mind would contemplate in full detail the structures of sticks, stones, stars and all other material objects, and that its complexly structured thoughts about them would actually be those material objects. In order to count as a material object, all that anything really needs is a structure of the sort physicists describe. You do not have to specify that this structure is carried by 'stuff of an essentially non-mental sort', any more than you need specify that something which is 'really you' cannot be made of the stuff of divine thinking. Sticks and stones and you and I, and absolutely everything familiar to us, might exist inside a divine mind as believed by Spinoza and other pantheists.

What is more, a divine mind could contemplate up to infinitely many worlds of kinds that could not contain conscious beings. The fact that our own world contains them could then illustrate an observational selection effect of the type examined in my *Universes* (1989). This included much discussion of Brandon Carter's *anthropic principle* that observers, intelligent living beings, must always observe themselves to be in intelligent-life-permitting situations.

3. I now take the idea of immortality much more seriously than before. Surviving bodily death could involve considerable 'disorder', in the sense that physical laws suddenly broke down, but *what it would be like to survive bodily death* might still be very much worth knowing. Also, it can be argued that a divine mind as imagined by pantheists, when it had 'thought all the way through' somebody's life on Earth, coming to the stage at which physical laws dictated that this life should cease, would have a duty not to 'switch off' the life by stopping thinking about it—like a scientist who, after developing a fully conscious machine, removes its electricity supply. So I might have grounds for quite expecting to survive bodily death, perhaps then joining with other survivors in learning more and more of the wonders of divine knowledge.

4. The book has much discussion of how, inside the wider system of a unified cosmic mind, various elements could be specially closely united. In particular, your own consciousness at any given moment may have a unity on which physicists could throw light. Quantum theory tells us that particles, even when they at first seem completely distinct, can be so closely interrelated that it makes little sense to treat them as *each separate in its existence*. Inside brains, such particles might combine to form unified wholes of startling intricacy. Ordinary computers might never be able to replicate the kind of unified experience of a complicated scene that is enjoyed by any baby and possibly by tadpoles as well. However, what are called 'quantum computers', now just starting to be developed, could perhaps soon do so.

It isn't obvious whether unity of the required sort could ever be had except in a world 'made of mental stuff'. In fact, it isn't even clear whether anything but mental stuff could ever exist at all. Maybe structures that are never in any sense experienced are too abstract to exist, as Bishop Berkeley thought, despite his failure to prove it. Very little of philosophical interest seems to me provable, which may help explain why this book of mine brings forward such inconclusive arguments.

The book owes an immense amount to many people. Besides feeling grateful to Mackie, I have Derek Parfit to thank for having drawn my ideas to Mackie's attention and for giving his own influential support to them later (although I cannot claim to have made an actual convert of him). I further owe a great deal to Ian Crombie, Ronald Hepburn, Hugo Meynell, Terence Penelhum, and Jack Smart. Also to A. C. Ewing for showing that an expert at philosophical analysis really can defend Platonism's creation story, which he applied to explaining God's necessary existence; to Nicholas Rescher for developing a rather similar story in his *The Riddle of Existence*; to John Polkinghorne for declaring in *The Faith of a Physicist* (the book coming from his Gifford Lectures) that the idea that *a divine being's ethical requiredness is responsible for that being's existence* may well be what is involved in the traditional theory that God is 'self-subsistent perfection' in which cause and effect are identified; and to Peter Forrest and Keith Ward for saying similar things in their writings. To Hugh Rice who held in his *God and Goodness* that divine creation of the world is probably best understood as 'just a matter of the world's existing because it is good that it should'. To Mark Wynn, whose book with the same title as Rice's leans towards the same theory. To Stephen Clark, Ivor Leclerc, and Hilary Putnam, three further writers of today who have defended similar views. To Timothy Sprigge for his insistence, echoed by Thomas Nagel, that there is a deep problem in how anything is ever conscious in a fashion making it true not just that it processes information successfully, but that there is 'something which it is like to be it': some way it feels to be it. To Peter Forrest (once again) and Leslie Armour, for joining Sprigge in keeping in print (and in forms apt to win the respect of philosophers in the contemporary British/North American/Australasian analytical tradition) views of the pantheistic kind I struggle to defend. To D. C. Williams for suggesting that, if time is as Einstein pictured it, then pitying the dead for *existing entirely in the past* makes as little sense as pitying someone for existing today but entirely at one location. To Bill Craig and Richard Swinburne who demonstrate how much can be done when you discuss God in an adventurous fashion, and to Robert Nozick for showing the same thing about philosophical discussions in general. To Brandon Carter for his ideas about *cosmic fine tuning* and *observational selection* which kept me occupied for many years. To David Lewis, the cosmologists Richard Gott and Werner Israel, and the mathematician Jean-Paul Delahaye for helping me to get people interested in Carter's argument that observers should, all else being equal, expect to find themselves in spatiotemporal locations where living beings of their type are most numerous, so that we should be suspicious of the idea, say, that only one in a billion humans will have been alive when we were. (If instead of spreading throughout its galaxy the human race became extinct shortly then, thanks to the recent population explosion, the figure would be nearer to one in ten, and this could be considered more credible. We can believe in God while still thinking our world a dangerous place.) To Lewis for something else as well: his famous defence of a 'modal realist' cosmos that may actually contain far more than my cosmos does. To Bernard Carr, Paul Davies, Jacques Demaret, George Ellis, John Polkinghorne, and Martin Rees, for advice and support in the field of cosmology. To Michael Lockwood, together with Ian Marshall, Roger Penrose, Abner Shimony, and many others, for applying ideas from quantum theory to the remarkable unity we find in our conscious states at particular moments, a unity which could be crucial to the sort of conscious life that is worth living. To the British Academy for inviting a lecture ('Our Place in the Cosmos') during my tour as the Royal Society of Canada—British Academy Exchange Lecturer for 1998, and to Anthony O'Hear of the Royal Institute of Philosophy for asking for another ('The Divine Mind') during the same tour. To the editors and referees of journals, collections of conference papers, book series, encyclopaedias, and so forth: in particular to Nicholas Rescher for choosing *Value and Existence* for his American Philosophical Quarterly Library, and to Routledge, Macmillan, and Prometheus for bringing out later books of mine. To everyone involved in publishing this present book and

Preface

especially to Peter Momtchiloff, Editor for Philosophy, to Charlotte
Jenkins and Jane Robson, Assistant Editor and Copy-Editor, and to
the experts consulted by Oxford University Press who made helpful
suggestions for improving it. To my wife Jill above all.

J.L.

Contents

Infinite Thinking   1

If God is real, then why are our lives so limited, so inadequate? Why is there anything except infinitely rich thinking, knowledge of absolutely everything worth knowing?

A possible reply is that nothing exists apart from the thoughts of God: infinitely many thoughts about everything worth knowing. Limited and inadequate though our lives are, they are still worth living, worth knowing about. An infinite divine mind includes full knowledge of how it feels to be living such lives, and this knowledge is the lives themselves. Their only reality lies in the fact that God is thinking them.

Imagine an infinitely complex mental life divided into regions: thoughts, that is to say, about separate groups of facts. Imagine a region filled with immensely much knowledge, all of it appreciated 'in a single glance'. In any region of this kind there could be no full knowledge of what you and I know, which includes precisely how it feels to be greatly limited and deeply ignorant. But the divine knowledge would presumably extend to that. Knowing everything in the least worth knowing, how could God be unaware of it? It could be known as more than just a possibility, for God's knowing it in all its structural detail could be its reality. As Spinoza saw, all our limitations and ignorance cannot refute the theory that you and I are tiny regions inside the divine thinking. (Could an infinite divine mind have full knowledge of how it feels to be atheistic? Why ever not? People who were elements in the divine being would not be thrown out of it through becoming atheists.)

(1) The chapter discusses whether infinite thinking is truly possible. The thoughts of a divine mind might be infinite even if, as suggested by Cantor, no mind could know the set of all truths 'because there is no such set'. (2) Thought about any complex structure would itself possess a complex structure, and a material universe could be fully real just through possessing a complex structure of the correct kind, the law-controlled kind that physicists investigate. It would not need to be made of 'the right sort of stuff', such as 'non-mental stuff'. Having the right structure would be sufficient. Among the divine thoughts there might be ones which combined to form many very intricate groups, each group structured in a way that justified our calling it 'a universe'. The divine mind could include infinitely many such universes. (3) It would no doubt include thoughts, as well, that were not organized into universes. Yet the divine thinking, although infinitely rich, might still not extend to many truths. Among truths about possibilities, many might not be worth thinking about, for instance because the possibilities in question were as disorderly as books filled with random letters.

The suggestion that reality consists not just of one infinite mind but of infinitely many, each worth calling 'divine', is considered only in later chapters.

A Pantheistic Approach to the Problem of Evil

If theism is correct—if God is a reality—then we face the theological Problem of Evil. Assuming that God is even moderately good, why are our lives so unsatisfactory?

The book will examine the kind of pantheistic answer suggested by Spinoza (1632–77). While his works are difficult, a natural reading of them is this. He views our conscious states *as elements in the thoughts of a divine mind that includes all reality.* Agreed, they could be rather inferior elements, but their details are still worth thinking about. Now, the divine mind knows or thinks everything worth knowing or thinking.

We can investigate these pantheistic ideas without worrying much about Spinoza's complicated writings. Did he consider that what was worth knowing included absolutely all truths about logical possibilities (matters describable without actual contradiction)? Did he picture God as contemplating abstract mathematical facts of fantastic complexity, together with all possible feeble jokes, bad poems, wicked actions, and depths of misery, or would he have said instead that everything in the divine mind had to fall into a single system ruled by what scientists call laws of nature? In his view, what was the logical status of worlds obeying laws different from the laws of our world? Were they each as inconsistent as a round square or an unmarried wife? Often extremely hard to answer, such questions could be of great interest to Spinoza scholarship yet we need feel no duty to answer them.

Instead let us see whether Spinoza's two crucial suggestions—that our conscious states are simply elements in a divine mind, and that it is a mind which knows or thinks everything worth knowing or thinking—could make the Problem of Evil more manageable.

Perhaps our world strikes you as very disappointing. Maybe you can imagine some other world you would much prefer to inhabit. But God perhaps thinks of that other world as well as thinking of ours. It might actually be a world containing somebody very much like you: perhaps even a person whose early years were like yours in every detail, that person's world and our world taking on different characteristics only later. Saying *you would prefer to be this other person* can have rather an odd status. To begin with, you could hardly use it as a complaint against the actual scheme of things, if a Spinozistic approach were right, unless you thought of your own conscious life as not worth having. For if yours is a conscious life worth having, and if the fact of your having it is just the fact of God's thinking of all your various conscious states, then presumably you should be glad that God is thinking of them instead of thinking of the life of that other person much like you, in that other world, without thinking of your life *also*. Besides which, how much sense could there be in wishing you had the life of the other person? Would this be any different from wishing you had the privilege of not being *you* at all?

Next, would you prefer to have divine thought-patterns in addition to your own so that you knew everything God knows? Would this make any sense? Perhaps not. It is hard to see how you could suddenly be given all the divine knowledge without ceasing to be you. Wouldn't sudden omniscience shatter your personal identity still more thoroughly than suddenly becoming a goldfish with its extremely limited thoughts? Essential to your being you, it might well be argued, is that your thoughts could extend only to a tiny part of what God knows. But the Spinozistic theory, remember, is that tiny elements in the divine thinking are what they are. So the Problem of Evil, if it is to have much bite, may have to depend on saying such things as that your own conscious life is simply not worth living. Now, would you go quite so far as to say that?

Trying to introduce ideas like these in the twenty-first century and in the West, and particularly to philosophers in the analytical tradition in which I was trained, one never knows where to start. The points I want to make could seem entirely natural to a traditionally educated Hindu, or to Hegelians such as F. H. Bradley, who sometimes called himself a pantheist but managed to reign supreme in Britain's philosophical world right into the early twentieth century, or to a physicist such as David Bohm, who speculated that all the parts of our universe form a collective mind of some sort; yet they can easily be dismissed as preposterous, for all kinds of powerful reasons. And it is no use thrusting a hand into the bucketful of possible objections, pulling out one of them and writing a book about it before tackling the next. Instead one has to paint a huge picture at speed, conscious that every brushstroke can earn raised eyebrows, incredulous stares, or worse. One has to do this because the elements in the picture make sense only when seen as a whole. From which it follows, unfortunately, that whatever one begins with can look outlandish.

The book's chapters are prefaced by chapter summaries. A thing to notice is that many of the themes mentioned in them are ones that feature in the writings of scientists. In particular:

Chapter 2, 'Minds Human, Artificial, and Divine', discusses the dramatic degree of unification that Spinoza attributes to our world.

Well, it is something that *quantum physicists* often say they have found in it. Again, Spinoza's pantheistic idea that the reality of everything is a matter of consciousness (for at least as I interpret him, he views the world in all its intricacy as nothing but intricate divine thought, divine consciousness) by no means forces him to believe that trees and rocks are conscious beings, and is actually something towards which quantum physics is fairly friendly. Many physical systems, quantum theorists have discovered, possess what Descartes viewed as the exclusively mental property of being more than just the sum of many separately existing parts. Even to predict the probable whereabouts of two photons in the same quantum state, you may need to appreciate that their identities are partially fused. While the chapter rejects the Cartesian idea of an immaterial soul, it points out that *quantum computers*—the principles governing them have already been demonstrated in laboratories—would work in ways that couldn't readily be imitated by any collection of cogwheels or transistors, or of atoms as conceived by nineteenth-century physics. They are ways in which brains, too, may operate.

Chapter 3, 'Time and Immortality', argues for an Einsteinian approach to the nature of time. Such an approach encourages the theory that divine thoughts about our world's events, and hence also these events themselves if (as pantheism suggests) their highly complex patterns are only patterns of divine thinking, *are all in some acceptable sense 'eternal'*, the world being 'unchanging' in a sense corresponding to this. This would not be denying that trains move and that children grow taller over the years.

Chapter 4, 'The Best and Infinity', insists that pantheism doesn't tell us we are powerless to influence the world's events. Instead of just waiting to see what the future will bring we can set out to make the world better, because all the causal patterns recognized by scientists can be found inside pantheism's cosmos. They are patterns which you and I can influence because our choices and actions *form part of them*. Scientists and philosophers have long held that this point is unaffected by whether physical laws govern the details of those choices and actions.

Chapter 5, 'Necessary Divine Existence', defends a Platonic creation story. It might at first seem one which scientists ought to reject.

Setting out to answer 'why there actually is Something, not Nothing', it appeals to *the ethical requiredness of there being Something*. Yet when you examine the matter closely you may well join the many scientists who think that the sheer fact of there being a cosmos—any reality whose laws science could investigate—isn't itself an affair that science can answer. Again, the apparent *fine tuning of our universe* might best be explained Platonically. Physicists and cosmologists talk of 'fine tuning' because many matters basic to the structure of the physical world, for instance the strengths of physical forces like electromagnetism and gravity, appear such that tiny changes in them would have prevented life's evolution.

All the same, theism looks quaint and outdated to many people today, while pantheism strikes many of them (and many theists also) as quite extravagantly bizarre. And if I saw no force in the Platonic explanation of why there actually is Something, something other than mere facts about possibilities, then—though some highly intelligent people could disagree with this reaction of mine—theism in general would seem odd to me as well, and I'd not rush to defend pantheism. As things stand, my confidence in the Platonic explanation is only a little above 50 per cent. It seems to me really quite likely that the world exists for no reason whatever. It could very well be that not even an existent of a supremely good type, a divine mind knowing everything worth knowing, existed *because of its ethical requiredness*. The mere sense of the words 'ethically required' cannot show that any ethical requirement, even the very strongest, will have any tendency to put itself into effect in the fashion that Plato envisaged when he wrote that the Form of the Good is what gives existence to things. Still, this Platonic theory is not grounded on any mistake in logic. It has come to be viewed by various very competent thinkers—they include leading philosophers in the analytical tradition of Britain, North America, and Australasia, and theologians of the kind who believe in supporting their faith with detailed arguments—as something which might actually be right. Yet surely it couldn't be right if the Problem of Evil couldn't be solved, and I cannot myself see how anything but pantheism would solve it. So the situation takes the following uncomfortable form, I suggest. Because you were unfamiliar with the Platonic explanation for the world's existence, the pantheistic world-picture of this book's first four chapters might well seem to you fantastic. In that case, maybe you ought to begin by reading Chapter 5. But what if you did? The Platonic explanation could then itself strike you as fantastic because you couldn't see how the Problem of Evil could be solved, and I'd wish you had instead begun with Chapters 1 to 4. There is just no escaping the fact that whatever one starts with can appear absurd.

As good a way as any of starting might be this, however. Let us ask what any Spinozistic pantheism could mean by *divine thought or knowledge*.

Divine Knowledge is Eternal Thinking, of Immense Complexity

Somebody can think something without knowing it, since people are often wrong. Again, one can know something without thinking about it. You knew all through the last five minutes that you weren't a purple cactus on Mars, didn't you? 'Thought', 'knowledge', and 'consciousness' are separate notions, and so is 'mind'. Our minds are not just collections of thoughts. They are what have our thoughts, before which they need to go through the process of generating them.

The case of divine thinking is supposedly very different. Here, thought and consciousness and knowledge and mind are rolled into one. Instead of struggling to generate its immensely many thoughts, the divine mind is in eternal possession of every one of them. They are all items of knowledge, and the knowledge is all of it fully conscious (unlike your earlier knowledge of not being the cactus). Also, the mind that has the divine thoughts can be merely the thoughts themselves, forming a unified whole: a whole in which they are united in their very existence despite there being infinitely many of them. This could be strongly linked to the fact that all of them are elements in the divine consciousness, because consciousness might be the only thing that can 'hold a many in one, a diversity within a unity', as Bradley expressed it.

A great deal of this is both controversial and obscure. For the moment, please just remember that the divine mind is being pictured as no ordinary mind in many respects, quite apart from the fact that it knows infinitely much.

Structures in the Divine Mind can be the Structures of Real Things such as Humans

On my Spinozistic or pantheistic theory, the structures of galaxies, planets, and continents, of mice and of elephants, and of you and me, as well as of the houses, fields, and streams with which we interact, are nothing but the structures of various thoughts in the divine mind. The divine mind does not contemplate any universe that exists outside it. Its thinking about our universe is what our universe *is*. When God contemplates various physical possibilities in full detail they do not remain 'merely possible' like the golden mountains of our dreams. They are genuinely real, existent, actualized, non-fictitious.

How, after all, do physicists describe material objects? Not with such useless phrases as 'good, solid stuff', but by trying to specify their intricate structures. In the divine mind those structures, including the structures which are the physicists and their laboratory equipment, are present in their entirety. If you consider this far from sufficient to make them into the structures of real stars, animals, pastures, atoms, electron microscopes, and scientists, then perhaps that simply shows how unsympathetic you are towards a Spinozistic world-view. For if, coming to accept such a world-view, you still protested that objects would be in need of far more in order to qualify as 'real', then you would have to declare that you yourself had no reality, which would be absurd.

You ought surely instead to continue counting yourself not only as real, but as a real material thing. Suppose we lived in a universe which a deity had created outside himself. It might then be appropriate to say that any structures in the deity's mind when it contemplated the things of that universe were 'mere models of those things, not real material things like us'. The Spinozistic theory, however, is that nothing exists apart from divine thinking, you and I being structures inside it. Now, it would cause endless confusion if those who accepted this theory went around declaring that they or other people were 'mere models and not real people' or that the material of the trees they had apparently bumped into 'wasn't really there'. Yes, when we humans imagine trees in great detail it would be asinine to claim that our minds thereby come to contain real trees. But the divine mind as a modern Spinozist conceived it would think of *absolutely all* the intricate structure of trees as described by a completely accurate physics. And the Spinozist would believe that, if granted a miraculously reliable vision of reality, then he or she, seeking elements whose structure was more or less as described by the best physics of the day, would find them in a divine mind whose structure was in whole or in part the structure of our universe, *and would find them nowhere else*. The Real would contain no other candidates for the description 'the material objects of our universe', and this point shouldn't simply be disregarded.

If you none the less want to say that Spinozistic pantheism 'makes the material world an illusion' because the material world as pictured by you and by most people, at least in the West, is something quite other than divine thinking, then so be it; use language as you please. But don't ask all Spinozists to define words exactly as you do, thereby forcing them to go around declaring that physical objects are illusory! The rules even of ordinary language are far from dictatorial when it comes to such matters as why, if at all, intricately structured material things couldn't be parts of intricately structured divine thought or consciousness—for since when have folk in the street been unable to communicate with one another without adopting firm views about what protons and electrons are made of, and how if at all they would differ from every element that a divine mind could contain?

Thought about Complex Structures Must itself be Complexly Structured, even when Divine

In science fiction you sometimes come across the idea that you are being deceived by a mad scientist's gigantic computer. Your friends have never really existed. There are only *simulations* which the computer is running so as to give you the illusion of interacting with other people in a world of good, solid stuff. You are a brain kept alive in a vat of nutrients and wired up to the computer.[1] A typical suggestion is that it would then be quite all right, say, to interact with the computer in the way that (if your brain is male and vicious) you had always thought of as *tormenting your wife*. No harm would be caused! She wouldn't be a real person, would she? Well, perhaps she would be. To have succeeded in deceiving you, the mad scientist's computer would presumably have needed to simulate people in immense structural detail. It is hard to see how the job could be done without the computer simulating everything right down to individual atoms—simulating them for itself, that is to say, so as to be able to generate data capable of deceiving you for years on end. The computer would, I think, have needed to build up inside itself a model containing elements corresponding to atoms, a model whose structure developed just as if those elements were indeed atoms interacting with one another. Regardless of whether the apparent wife should be called 'a genuine wife', it then becomes doubtful that no misery would be being given to a real person, a centre of thought and consciousness. Couldn't even a computer simulation have a consciousness of its own if its structure were intricate enough?

Possibly it would lack consciousness. Having the right kind of intricacy may not be enough to make a conscious being. Perhaps there has to be something further, namely, having states in which hugely many elements are united in their existence: states of a kind that some people think could never be present inside any computer, while others suggest that quantum computers might some day come to contain them. Whether the apparent wife was a conscious person might depend on something more than the degree of complexity of the computer simulations. It might depend on what kind of computer (a quantum computer, or just one made up of items like transistors interacting in the ordinary way) the mad scientist used.

Suppose, now, that you were tempted to dismiss any divine mind's thoughts about humans as 'mere simulations, not genuine humans'.

---

[1] The brain-in-a-vat scenario is discussed in Leslie 1989*d*. This argues that brains in vats could have complex thoughts, learn languages, and enjoy very full lives.

What could be your grounds? Would you consider the divine mind too limited to contemplate all the structural details that humans have? Or would you say that elements inside any such mind could never be unified, united in their existence, in the style in which the elements of human conscious states are unified? Either way, your grounds could be considered very weak. But alternatively, would you think a divine mind capable of contemplating complex things *without itself containing any complexity*? I have been taking it for granted that it is impossible for anything to be thought about in great structural detail without there being an equivalent richness of structure in the mind doing the thinking. People sometimes deny this, however.

I believe nothing quite so silly as that, whenever a man thinks of his house, his brain must contain a tiny three-dimensional house. I instead assume that people's thoughts about houses are seldom very detailed. I also assume that, even when a brain thinks of a three-dimensional structure in great detail, the correspondence between this structure and any pattern of nerve-cell firings (or whatever) inside that brain must be in many respects unlike the correspondence between a doll's house and a full-sized house. It could be rather more like the correspondence between a map and a countryside, or between the sounds of a symphony and the laser-readable marks on a compact disc. In terms of structure, however, it could still be a good correspondence.

What is meant by 'structure' here? You could start by asking mathematicians. Mathematics can give precision to the idea, for instance, that various surfaces of different shapes and sizes share the following structural characteristic: all of them are surfaces of rectangular blocks in Euclidean space. Again, mathematics can specify speedily and unambiguously an important respect in which successive electrical impulses entering a loudspeaker are structurally similar to successive soundwaves coming out of it. To specify this you wouldn't have to wait for any explanation of how the loudspeaker functioned—although you could of course need to give one if you wanted to convince people that the structural similarity as specified did deserve to be called 'important'. If in search, however, of something nearer to the case of a human mind thinking about this or that, then why not consult experts on artificial intelligence? Ask how their computers manage to model such things as the positions of kings, knights, and pawns during games of chess. Don't expect them to tell you that the computers keep track of various complicated situations on chessboards without 'containing models of them, elements forming similar structures', in any useful sense! Elements inside a computer can form models of chessboards and chessmen without being made of wood or plastic. A structure consisting of items arranged intricately in space can even be represented by one whose elements are ordered in time without ceasing to be *well modelled*: duplicated, that is to say, with respect to structural features that are important here. And parts of a perceived scene could be modelled inside a brain as being close together in space despite how neuronal activities corresponding to those parts were split between the cerebral hemispheres.

What if you consult theologians, though, or philosophers of religion? It turns out that many of them oppose the idea that God's mind, thinking about this or that intricately structured situation, has itself to be structured intricately. Aquinas is a prime example. True enough, you can find Aquinas saying that God 'sees things other than himself' by exploiting the fact that 'his essence contains the likenesses of other things': it 'takes up the form proper to plant', for instance, when God cognizes the fact that there can be the sort of life that plants have. God knows absolutely all other things just by knowing himself because he contains the likenesses of them all. Yet assertions like these, suggesting a divine mind of immense structural complexity, are combined with insistence that God does not understand things 'by composing and dividing'. He is Pure Being, which means he is not characterized by any complexity or by anything we would recognize as qualities. God's creative power, God's freedom, God's justice, God's mercy, God's goodness, God's knowledge, are all of them strictly identical with one another and with God's act of existence. To the amazement of many philosophers of today, Aquinas went so far as to believe that humans are loved by God without God himself standing to humans in the real relationship of loving them. The idea appears to be that, just as Mr Black could get to be less tall than Mr White while Mr White remained unaltered, the change being due entirely to shrinkage on the part of Mr Black, so a human could get to be loved by God without God himself being any different from how he would have been, had he decided never to create that human.

Ideas like this are by no means peculiar to Aquinas. They are defended by many Christian theologians and philosophers today, both Catholic and Protestant, and are found in contemporary Shi'ite thought as well. Aquinas never claimed to be able to make any clear sense of them, however. He ascribed this to how God was so incomparably great, so different from anything the divine power had created, yet it could instead simply be that they amount to nothing sensible. The theory of Spinozists like myself is rather different. We suggest that the elements in the divine mind are all so closely united that, like the mass, the shape, and the colour of a lump of cheese, they do not exist each in isolation from the others. While this may be a difficult theory, it is nowhere near as hard to understand as Aquinas's view that the divine mind doesn't truly have elements of any kind since it is totally lacking in structure. The right analogy for that could be a lump of cheese whose colour was identical to its shape.

In effect, Aquinas's position could be on a par with the Hypothesis of the Pure Ego which, until the early twentieth century, was admired for its alleged ability to explain how somebody could remain the very same person from one year to the next. As C. D. Broad expressed it, 'There is a single Pure Ego which lasts without qualitative change throughout my life and owns all my successive states' (1925: 279). The notion was that an ever-altering mental life could all of it be 'possessed' by something that could appreciate the alterations without itself altering. This remarkable entity could continually become aware of new mental states so that it could be, for example, sad at noon and happy at midnight, *without ever becoming any different* ('no qualitative change', remember). To me, that looks as flat a contradiction as you can get. Now even if, thanks to his doctrine that God is outside time, Aquinas avoided landing in this particular contradiction, he could seem to have ended up in one every bit as great when he taught that God's existence would necessarily be exactly the same no matter what went on in the world he had created (or even if he had chosen not to create it).

Aquinas could hardly have been more mistaken, I suspect. Instead of having supreme simplicity, in the sense of lacking all structure, all ordering of elements each at least partially differentiated from every other, a divine mind would carry an immensely complex pattern with innumerable elements. Still unconvinced? If so, then it is hard to see how the matter could ever be proved to your satisfaction. For some interesting and expert discussion, though, you could go to the writings of Alvin Plantinga, Keith Ward, W. L. Craig and W. P. Alston.[2] These contain helpful references to various pages of Aquinas, his admirers, and his critics.

The points I have just been making are not intended as an attack on the competing theory suggested by other pages of Aquinas, the Platonic or Neoplatonic theory that 'God' is the name not of any being but of a creative force whose power is inseparable from its goodness, a force in no need of guidance from any complexly structured mind. Aquinas is an undeniably great philosopher and this competing theory is far from nonsensical, as Chapter 5 will discuss. (He frequently insists that talk of God *is only analogical.* This can suggest that he viewed the competing theory as really no different from the theory I have been challenging, the theory that God could think about things with a mind that had no structure whatever.)

---

[2] Plantinga 1980; Ward 1996*a*: esp. 211–14, 229–30; and Craig 1999, all treat Aquinas's position as incoherent. Alston 1986, however, comes close to it, speaking (pp. 297–9) of 'immediate awareness' as being 'a direct and foolproof way of mirroring the reality to be known', but one in which the mirroring involves no 'mental maps' or 'inner representations' since here the state of knowledge 'is constituted by the presence of the thing known'. Yet even when such awareness is attributed to God, not humans, it is hard to see how Alston's position can be much different from the Pure Ego theory. How could any mind come to know the complex structure of anything *without itself taking on an equivalent complexity*? And what would taking on the complexity be, if not a case of forming some kind of *inner representation*? (It is no use arguing that an inner representation couldn't be of any help because it itself could be known only with the help of some further inner representation, and this in turn through the aid of another, and so on, in an infinite regress. You might almost as well argue that robots that form inner representations of their surroundings—as many of them now do, to help them to operate usefully—must have infinite regresses in their interiors.)

Up to Infinitely Many Universes Exist in the Divine Mind, Together with Many Things Not Organized into Universes

If the divine mind did have a complex structure, and one with as many elements as there were items in the divine knowledge, then how extensive would the structure be? The traditional doctrine is that the divine mind would know infinitely much. In *The City of God* (12. 18) Augustine writes that people who affirm 'that God cannot know things infinite' might just as well 'leap right into the pit of impiety by declaring that God's knowledge of numbers is limited'. He pictures God as eternally thinking not only of all possible numbers, but of absolutely all truths. Only wretches would dare to set limits to what God knows.

Two ideas are at work in Augustine's thought: first, that God's unchanging knowledge is infinite, and next that God knows absolutely everything. The second idea goes far beyond the first, at least if you accept such commonplace mathematical claims as that *the whole numbers are infinitely numerous* so that any mind knowing all of them 'would know infinitely much' (in some fully acceptable sense) even if largely ignorant about everything else. Suppose that God's eternal, unvarying mind contemplated nothing apart from how the ultimate constituents of various universes were ordered in space and in time. While perhaps then knowing infinitely much (for some of those universes might be infinitely large, or there might be infinitely many of them) God would none the less remain ignorant of all sorts of affairs—of hugely many mathematical facts, for example; of hugely many silly thoughts which might be had; of hugely many possible ways of causing torment. The divine knowledge would not extend to such matters even in the sense of being able to give answers if asked (much as you could have answered had anyone inquired whether you were a cactus). An eternal, unvarying divine mind isn't the sort of mind that could suddenly answer queries which it hadn't yet contemplated.

Let us defer considering whether the divine mind is aware of absolutely all truths about actual or possible situations. For the moment let us just imagine that among the things it contemplates is the entire structure of *a second possible universe* much like the possible universe in which you and I find ourselves. (Our actually existing universe is a possible universe. All actual things are possible ones *as well*.) Now if my earlier arguments were correct, what ought we to say about the divine mind's contemplation of the entire structure of this second possible universe? Might it be a case of knowing a structure which remains that of a merely possible universe, one having no actual existence? Not so, I suggested. A material universe, my suggestion ran, doesn't need to be composed of any particular variety of stuff, such as 'non-mental stuff'. All it needs is a structure of the sort physicists investigate. Well, my hypothesis is that the divine mind contemplates absolutely all the structure of the possible universe in question. And if my arguments have been on the right lines, then its contemplation of this structure would necessarily involve its having such a structure itself, either as a whole or else in some region of its being, because no mind, divine or otherwise, can think about a structure in its completeness without itself being equivalently structured either in part or as a whole. Doesn't it then follow that the divine mind would contain this second universe as one which, although 'made out of mental stuff', was still an actual material universe?

You might at this point protest that divine thoughts about the order of our own universe's constituents, if they could somehow be placed side by side with the constituents themselves 'which are obviously something different', would be found to be 'structured similarly' only in a sense somewhat like the one in which a symphony and various marks on a compact disc can be structured similarly. The parallelism, you might hold, could never be absolutely perfect, no matter what tricks omnipotence used in an attempt to make it perfect. And while some people could perhaps still want to say that when thinking about our universe the divine mind could produce so close a parallelism that it could be said to contain 'a material universe made out of mental stuff' although our material universe itself *wasn't* made out of mental stuff, might you not feel inclined to forbid this way of talking? Might you not demand that no universe made out of mental stuff should ever be called 'material' by anybody? 'It would actually be better', you could be tempted to insist, 'to maintain *that no material universe existed anywhere* than to let the words *"material universe made out of mental stuff"* pass our lips.'

Spinozists like me would ask you to resist the temptation. We maintain that even the universe that you and I inhabit is made out of mental stuff, the stuff of the divine mind, and that its structure is of the sort investigated by physicists which makes it a material universe.

What, after all, could we know about the ultimate stuff of our material universe and how its elements are related to one another, and about the ultimate nature of any divine thought or consciousness that there might be, and how *its* elements would be related to one another, which could assure us that nothing worth the name of a material object could be composed of elements that were elements of divine thought or consciousness? Until fairly recently almost all philosophers felt they knew that human minds couldn't conceivably be ingredients of the physical world 'because they had quite the wrong properties for this'. It would seem a curious failure to learn from past errors if, almost all of them having convinced themselves that this was mistaken, the descendants of these philosophers then proclaimed that material objects couldn't conceivably be ingredients of a world of divine thought 'because they would have quite the wrong properties'. For exactly why would the properties be wrong, please? Material objects, for instance trees and rocks, might exist inside pantheism's divine mind *without themselves being conscious, thinking things*. Compare Thomas Nagel's treatment of panpsychism, which he defines as the view that 'the basic physical constituents of the universe have mental properties, whether or not they are parts of living organisms'. Although finding panpsychism difficult to accept, Nagel writes that it 'appears to follow from a few simple premises, each of which is more plausible than its denial'. The premises are (1) that we are composed of matter that had a largely inanimate history; (2) that mental states like thought and feeling are neither physical properties of organisms nor implied by physical properties alone; (3) that they nevertheless are properties which we have as physical organisms, because we lack immaterial souls; and

(4) that all intrinsic properties of complex systems derive from the properties of their components. But while these apparently reasonable premises can result in panpsychism, they *do not*, Nagel rightly insists, 'entail panpsychism in the more familiar sense, according to which trees and flowers, and perhaps even rocks, lakes and blood cells have consciousness of a kind' (1979: ch. 13, 'Panpsychism'). Viewing everything as having mental qualities, or even as being 'entirely made out of mental stuff' (such as the stuff of the divine mind), does not mean you should start wondering whether it is cruel to crush a pebble or boil a potato. Developing his version of pantheism, Peter Forrest takes care to explain that he is suggesting 'not that all things have the property of being conscious but rather that all things have the property of there being consciousness of them' (1996: 203).

Pantheists can well believe that the divine mind carries, in addition to the structure of our universe, an immense amount of further structure. If it does carry it, then no doubt some of this further structure could be described as complex thoughts about all the things in this universe. There might be such divine thoughts as the following: that it is a universe of more than thirty trillion intelligent living beings. Or again: that if one wanted two examples of unusually unpleasant humans, Stalin and Genghis Khan would do nicely. Or (for who are we, as Augustine asks, to set limits to God's knowledge?) conceivably even this one: that the greatest number of ants ever trodden on by a dinosaur during a period of 5.46 minutes was 3,479,992, the fifth smallest of the ants weighing approximately 0.041 grams. But the fact would remain that besides containing such thoughts, each of them itself fairly complexly structured, the divine mind would include all of the immensely complex structure of our universe. At least as I am developing it, pantheism is the theory that being real inside that mind is the only reality that our universe has.

Pantheists of today can next find it natural to think that the divine mind carries the structure not just of the universe we inhabit, but of infinitely many others as well. Why should any pantheist imagine that God contemplates only a single universe when today's journals of theoretical physics and cosmology are filled with articles taking it almost for granted that universes exist in infinite number? As I explained in *Universes*,[3] people typically have two chief reasons for believing in universes beyond our own. The first is that, after dreaming up mechanisms which might operate at the coming into existence of our universe, they are reluctant to believe that such mechanisms operated on only a single occasion. The second is this. Suppose there existed hugely many universes and that they differed widely in ways for which technical explanations can be suggested. (Varying scalar fields, for example, could split apart the forces of nature in different ways in different universes.) The existence of these many and varied universes would solve a major puzzle. It would become unsurprising that there existed at least one universe, ours, in which everything was 'fine tuned' in a fashion encouraging the evolution of intelligent life. Well, Chapter 6 will say more about these two reasons, which can look very forceful; but whether or not they are, pantheists can add a third reason to them. It is that the divine mind, knowing all that was worth knowing, would surely know the intricate, beautiful structures of innumerable universes.

How Ignorance and Change can Exist Inside a Mind Vastly Knowledgeable and Eternal

While the divine mind would contemplate universes in all their details, you and I are ignorant of almost every detail even of our own planet. The Spinozistic suggestion is that we can be elements in the divine thoughts without ourselves experiencing anything like those thoughts in their infinite completeness.

Forrest is a practitioner of modern analytical philosophy who accepts this suggestion. 'Our minds', he writes,

are distinct from the divine mind just because our minds are integrated subsystems of the totality of things. But this distinction is the difference of the part from the whole, not the difference between two non-overlapping things. That is because our minds are parts of the content of the divine mind. The

---

[3] Leslie 1989*a*. Among several other items listed in the Bibliography, note particularly the edited volume of 1990, *Physical Cosmology and Philosophy* (reappearing in 1998, expanded, as *Modern Cosmology and Philosophy*).

divine awareness of the things you and I are aware of is numerically, not just qualitatively, identical to your or my awareness. (1996: 202)

His, he adds, is a theory 'in which God literally shares our joys and sorrows'. Because these joys and sorrows are one and the same as some of the divine mind's own experiences, just as Bonaparte was one and the same as Napoleon, there is no question of God's being ignorant of how these joys and sorrows feel to us.

Many theologians react strangely to any position like Forrest's. They deny that we could be parts of God's own reality, subsystems of thought having their own well integrated distinctness from the rest of that reality, and yet they claim that the divine knowledge extends to precisely how it feels to experience human sorrow, being confused, being in terror, and such things as the thrill of murdering a man for his wallet. Listen, for instance, to Oxford's Regius Professor of Divinity, Keith Ward, as he discusses God's awareness of a torturer's joy at torturing. God's experience of the torturer's feelings, he writes, 'would not be the experience as the torturer has it', but at the same time God knows (or at least, instead of saying with A. N. Whitehead that such feelings *are included in* God it is 'probably better to say' that God knows) 'exactly what it is like to have them', although in a manner 'wholly unparalleled in human knowledge' (Ward 1996a: 251, quoting Whitehead 1938: 350). This can certainly look like trying to believe the impossible. In any region of the divine being that was flooded with love, how could there be knowledge of precisely how it felt to be Stalin or Genghis Khan? And how about complete awareness of what it is like to be utterly terrified? How could this be fully fused with consciousness of being God, with nothing to fear? How, again, could a mind know just how it felt to be deeply ignorant when all of it was vibrant with awareness of knowing everything worth knowing? The next chapter will return to this area but what we ought to conclude could seem plain. Can a divine mind have experiences with a flavour exactly like that of our own experiences? Yes, but only if, inside its richly structured thoughts, some well individuated subsystems *are in fact our experiences.* Our mental processes are, I think, brain processes, and the divine mind, knowing everything worth knowing, knows all about brains at the level of their constituent quarks, leptons, or yet tinier components. Still, it is not at that level alone that it knows brains. It must also know our brain states in the largely ignorant ways in which we know them, for otherwise it could not know *exactly how it feels to us* to be in this or that mental state. It must know what is going on in our brains in the way that we ourselves ordinarily know it—and we certainly do not know all about the quarks and the leptons, or even about individual nerve cells. For reasons on which quantum theory may shed light, we can sometimes grasp as fully unified wholes various complicated cerebral realities, but this isn't to say we know all about trillions of quarks.

How about *our experience of time's passage*? I hold with Timothy Sprigge—who has kept Spinozistic ideas alive in Scotland just as Forrest has in Australia—not only that our world consists 'of innumerable finite centres of experience', centres all of them united in a consciousness cosmic or divine, but also that there is a sense in which the experiences enjoyed by these centres 'are all just eternally there'.[4] How can this avoid being nonsense? How can the theory that all experiences are ultimately parts of an unvarying divine mind be reconciled with the plain fact that our experiences are constantly altering? (Given that he is about to tell us that God and the cosmos are one and the same, God having all things inside himself, what can Spinoza mean by writing in the opening chapter of his *Short Treatise on God, Man, and His Well-Being* that because of being perfect God 'cannot change into anything better' and must therefore be 'immutable'?) Chapter 3 will discuss the matter at some length. For the moment, let me quickly answer that the world, even if all of it is 'eternally there', might still include realities worth the names of 'time' and 'change' *if those words were given appropriate senses*.

They would not be the sole respectable senses. A situation that is eternally there can be thought of as unvarying against a background of possible changes and therefore as being 'frozen in time', 'without any change', on one viable understanding of the words 'time' and

---

[4] Sprigge 1997: 202–3. After distinguishing his own carefully described position from others that also use this name, Sprigge calls it 'pantheism'.

'change'; however, a second interpretation of the words could be equally viable. Which interpretation you selected could be simply a matter of preference. What we need here, I suggest, is the moral Albert Einstein drew from his relativistic formulas: that we should be thinking in terms of 'a four-dimensional existence instead of, as hitherto, the evolution of a three-dimensional existence' (Einstein 1962: appendix 5, p. 150). But once having accepted Einstein's idea, how are we to *talk about* the world? Can we recognize any realities reasonably describable as 'replacement of some situations by others' and 'the passage of the years'? Certainly, yet only so long as we take this to mean merely that different cross-sections of our four-dimensionally existing world do differ in their characteristics. People of the eighteenth century are absent from the twenty-first, but in a way strongly analogous to that in which people in Toronto are absent from Vancouver. Acorns do develop into oaks but only in much the manner of a road starting off narrow in Berlin and developing into something broad before reaching Paris. Sure enough, we experience changes in our successive states of mind—yet not changes of quite the kind typically imagined by the man in the street, because no situation ever gains existence absolutely and then loses it absolutely. So if we instead choose to define 'time' and 'change' as involving existence absolutely gained and later absolutely lost, then Einstein's world is timeless and changeless. There is nothing wrong in choosing to define 'time' and 'change' in that other fashion.

The key to understanding this area is that Einstein's beliefs are in fairly clear conflict with common sense. Accepting Einstein's worldview, we cannot keep everything that the man in the street will tend to see as implied by talk of time and of change. We have to throw something out while keeping the rest. Just what we are to throw out is an arbitrary affair, so that no one way of talking about the area is 'right' in a manner that makes the contrasting way 'wrong'.

According to the Einsteinian position—popular among philosophers thanks in particular to the writings of J. J. C. Smart and Adolf Grünbaum—'now' is best treated as a word behaving like 'here'. What is *here* for me can be *over there* for you, and what is now (or 'in the present') for us can be in the future for those who are (relative to us) dead, but who have their own *nows*. While Einstein had interestingly strong grounds for accepting this, it seems to me impossible to prove it; but disproving it is equally impossible. No simple appeal to everyday experience can do the trick because Einstein is not denying that we engage in constant struggles, often in radical ignorance of what the next moment will bring. The heard thunder of a train, the felt gust of its passing, the dizzying sights and movements of a switchback ride, the feeling of being confined to the present and able to peer beyond it only through the dusty lenses of memory and prediction, can all of them be there in the Einsteinian picture, because *What patterns do experiences bear at particular dates?* must not be confused with the philosophical question, *Is existence transferred from each pattern to the next, or are presentness, pastness and futurity only relative?*[5]

Problems with Divine Knowledge of Absolutely Everything

We now come to what can well seem a very major problem. Is knowledge of all truths possible, and if so, would it be desirable? Were it both possible and desirable, then we might have to abandon pantheism of my kind because it might tell us to reject inductive reasoning. It might, in other words, instruct us to expect our experiences to become disorderly at any moment. Rational people could never form any such expectation. Yet we could be forced to form it when pantheistic ideas became combined with the theory that God, contemplating all that is worth contemplating, knows the structures not just of all orderly worlds, but also of all possible scenes of disorder.

As already mentioned, pantheists of today can very naturally believe that the divine knowledge extends not simply to our universe but to hugely many other universes as well. Now, imagine that God has full knowledge of the structures of all possible universes that obey anything worth the name of causal laws, universes characterized by

---

[5] I am quoting from Leslie 1979: ch. 9, which discusses time. For Smart's views, perhaps see Smart 1967 or 1989: ch. 2; for Grünbaum's, perhaps Grünbaum 1967 or 1973.

regularities such as scientists investigate, so that all these universes actually exist inside the divine mind, exactly as ours does. This raises no obvious problem for trust in induction, that is, for confidence that the future will resemble the past in ways we could hope to understand and to exploit. Inductive reasoning will be a useful guide in a very large range of causal-law-obeying universes, so why not in ours? Yet what if we next suppose that, because the divine knowledge covers *absolutely everything,* God further knows all about hugely many universes that do not obey causal laws, universes as disorderly as you could dream of? Once we grant that the divine knowledge extends to absolutely all truths, and therefore to all the details of all possible universes, how can we deny that it extends to the disorderly universes as well? But now comes the difficulty for pantheists like me. Mustn't God also know all the details of universes which start off orderly *and then become disorderly*? But if so, then why should we imagine that our own universe is not of this variety?

Remember, my Spinozistic theory is that, whenever God contemplates a possible universe in all its details, then an immensely complex structure has to be present in God's mind. The fact that the structure is 'made of mental stuff' is no adequate ground for calling it unreal. Our universe is just a structure in the divine understanding, and so are all other universes that there may be. Well, our universe appears to have developed in orderly ways up to date. But if God contemplates all possible universes in all their details, and if all of them are therefore fully real universes, ones which do indeed exist, then it would seem to follow that there exist hugely many universes which, after developing exactly as ours has done until the present moment, are due to become disorderly at the very next moment. Now, why fancy that we aren't in one of those universes?

David Lewis faces the same problem when defending his well-known *modal realism*. This is the philosophical theory that each and every possible world is really existent—the term 'world' meaning not an inhabited planet but an entire connected scheme of things. Not all of the worlds are what Lewis calls 'actual' but this is because he asks each intelligent living being to apply the term 'actual' only to the world which he, she or it inhabits. Compare how 'here' is a term applied only to what is local. Saying that various things are *here* doesn't deny that many others exist *elsewhere*: in the next house, for instance, instead of in this one. Similarly, when he says that our universe *is actual* while other possible universes *are not actual,* Lewis is merely recognizing that the other universes aren't where we exist. In his opinion all of them exist somewhere. All are actual to any intelligent living beings that inhabit them.

Lewis would therefore need to believe in everything in which I believed, just so long as my beliefs extended to nothing that was utterly impossible. If a divine mind knowing everything worth knowing is a possible mind, then Lewis ought to believe in it. On his theory the Greek gods must themselves exist somewhere, assuming (and why not?) that they are at least logical possibilities. 'I am perhaps the most extreme polytheist going', he writes, explaining that he does not consider that a being has to 'satisfy some inconsistent description to be a god'. He accepts endlessly many gods despite picturing our own universe as 'entirely godless' (Lewis 1983: p. xi). Now, the divine mind of my pantheism could be judged no more impossible than his Greek and other deities. The result is that, even granted that he believes in universes that aren't parts of divine minds, it could be difficult for him to find grounds for thinking that he himself exists in one of those universes instead of inside a divine mind. In my eyes, though, this would be no argument against his modal realism.

Lewis is correct in writing that 'incredulous stares' do nothing to refute him. Yes, his theory does (as he at once goes on to say) tell you that there are 'uncountable infinities of donkeys and protons and puddles and stars, and of planets very like Earth, and of cities very like Melbourne, and of people very like yourself', yet this does not put him on a definite collision course with the demands of simplicity. As has been forcefully argued by R. H. Kane, by Robert Nozick, and by Peter Unger, a cosmos in which absolutely all possibilities were actualized might actually be considered *simpler*—just look at how few words ('a cosmos in which all possibilities are actualized') would be needed for identifying it!—than a cosmos that included merely some of them (Lewis 1986: 133; Kane 1976; Nozick 1981: 123–30; Unger 1984).

Besides, reality-as-a-whole couldn't be made *vastly richer* because

25

Lewis was right. Whether or not Lewis is right, reality-as-a-whole is infinitely rich. You need not share my taste for pantheism in order to accept this. The Real is infinitely rich because infinitely many possibilities, some of them possibilities which are themselves infinitely intricate (for instance, possible universes each extending infinitely in space and in time), *really are* possibilities. Lewis may be wrong when he classifies them as real existents, every one of them. Many may be nothing more than real possibilities. But a possibility is no less real, and no less intricate, simply through remaining confined to the land of the possible. The kingdom of possibilities isn't a fiction. If it were, then there would have been no conceivable alternative to whatever actually exists, which is a doctrine few would willingly accept.

Where Lewis can nevertheless appear in genuine trouble is over induction, the method of reasoning that assumes that the future will obey the same basic laws as the past. He expresses the point as follows:

According to my modal realism, there are countless unfortunates just like ourselves who rely on reasonable inductive methods and are sorely deceived. Not the best but the third best explanation of their total evidence is the true one; or all their newly examined emeralds turn out to be blue; or one dark day their sun fails to rise. To be sure, these victims of inductive error differ from us in that they are not actual. But I consider that no great difference. They are not our worldmates, but they do not differ from us in kind.'[6]

In view of this, may we not have to conclude that (as he puts it) any modal realist 'has no right to trust induction—he should turn sceptic forthwith'?

Lewis offers us two replies. The first is that everybody else is in the same boat as he is. It is 'possible, and possible in ever so many ways, that induction will deceive', but everyone has to accept this. By trusting that our world won't suddenly start behaving crazily 'we run a risk'; but people cannot deny this, whether or not they believe that worlds of all possible varieties exist in parallel to ours.

Is that an adequate reply? If so, I would recruit it to defend my pantheism, but unfortunately it runs into the following problem. For every

[6] Lewis 1983: 23. All the immediately following quotations are from this page, the next, and Lewis 1986: 115–23.

way in which a world could continue in an orderly fashion, there can appear to be many more ways in which it might continue in a disorderly fashion. Suppose we are watching a stone as it starts to fall at a time when no winds are blowing. Lewis and I agree there are countless ways in which stones exactly like this one could conceivably behave in universes that had all of them been exactly like ours right up to the present moment. Any such stone could fall in a straight line slanting slightly to the north, or a lot to the north, or by a medium amount to the north-east, etc.; or it could fall in a spiral or in a zigzag in which each zag was 2.3 times as long as the preceding zig, or in another zigzag in which the figure was 55.9; or it could stop falling and hover in mid air, or explode, vanish, or turn into syrup. What if these and all other possible forms of behaviour were each adopted somewhere, in worlds that had until now developed in precisely the way in which ours had done? It could then seem almost certain that you and I, if continuing to watch the stone instead of ourselves vanishing or becoming syrup, would find ourselves in one of the worlds in which it behaved in a style that no rational person would expect.

What is our actual situation, though? Sure enough, you and I may have to accept that hugely many *possible people* would find themselves in worlds that started to develop crazily, yet at least we can believe that those possible people are merely possible, there being no really existing world in which stones behave in ways so fantastic. Lewis, in contrast, could seem forced to defend the position that worlds in which stones suddenly act bizarrely not only exist but are much more common than worlds in which they continue to behave as you and I would rationally expect. And then instead of simply being in the same boat as us, the boat of not being able to guarantee that inductive reasoning will deliver the right results, he could seem to be doing his best to guarantee that it would deliver the wrong ones.

Lewis, however, is an extremely skilful philosopher, and his second reply hurries to the rescue of the first. It is a flat denial that the possible worlds in which inductive reasoning failed would be in the majority. He argues that the worlds in which it failed *and also those in which it succeeded* would be infinitely numerous, and that in consequence nobody ought to claim that it would fail 'in most of them'.

Look again at my stone. Of the conceivable ways in which this stone could move, infinitely many would be only barely different— quite undetectably different—from falling precisely vertically all the way to the ground. How, then, could anyone be forced to accept that 'most' ways of falling would make inductive reasoning fail?

To drive home his point, Lewis considers the case of the prime numbers. We might be tempted to claim that these were obviously extremely rare among all the whole numbers because the proportion of primes gets smaller and smaller as you continue counting. But wait! Aren't we dealing with infinities in each case? Cannot we therefore challenge any particular 'partitioning' of the primes and the non-primes? We might, for instance, challenge the partitioning that would be set up by starting to count from the number one, collecting the non-primes into four columns which soon started to lengthen far more rapidly than the single column into which we put every new prime. Another partitioning of the primes and the non-primes would reverse this situation.

Similar reasoning attracted Georg Cantor whose approach will be discussed in the next section. It has much in its favour. Experts often do say such things as that talking of 'the low probability that a whole number picked at random from the infinitely many whole numbers will turn out to be prime' is nonsense. But other experts, particularly among the physicists, reckon that there can be 'natural' methods of partitioning infinite sets, and that science and common sense in effect rely on these methods. Suppose we grant that the number of points on a dartboard is strictly infinite both outside the tiny central bull's-eye and inside it. It can still make sense to say that the chances of a dart's hitting the bull's-eye are small because the set of points outside it is 'importantly larger' than the set of points inside. Maybe not larger *in number* since infinities, at least if they are at the same Cantorian level, cannot be said to differ in number, at least if Cantor's way of understanding 'equality in number' is accepted; but larger none the less. Or at any rate *different* in a fashion justifying, prior to actual experiments, the belief that darts thrown at a dartboard will mostly not land in the bull's-eye unless thrown by the specially trained. A standard way of expressing the difference is that *the range* of the points outside the bull's-eye is larger.

This can be important in quantum theory. Apparently absurd possibilities are envisioned here: for example, suddenly finding that the particles of a neutron star had all of them decided to jump inwards so that the star collapsed to form a black hole. One sees actual calculations of how long you would have to wait on average—it turns out to be a number of years so great that, writing it out by starting with a nine and then adding zeros, you would need many more zeros than there are atoms in our galaxy—in order to observe any given neutron star behaving in so strange a manner. Now, quantum theorists have confidence in such calculations even when thinking that infinities are involved. Yes, the number of ways in which the inwards jump of the particles could occur, and the number of ways in which it could fail to occur, may perhaps both be infinite, but the range of the ways in which it could occur is comparatively tiny.

Consider, too, the possibility that there exist infinitely many universes very much like ours, or that our own universe is itself infinite as can seem to be suggested by observations (which are often thought to reveal a universe whose gravity is too weak to wrap its space around until it joins up with itself like the finite surface of a sphere). We might then expect the existence of infinitely many planets much like Earth, and infinitely many cold puddles that suddenly began boiling through drawing heat from their equally cold surroundings. After all, elementary physics—no need to introduce quantum theory here!—tells us this is always possible, though it is normally considered not nearly as likely as tossing a trillion heads in a row. What if we argued that the infinitely many puddles which failed to boil couldn't in any fashion outweigh the infinitely many which did boil? Even if the merits of inductive reasoning impressed us enough to make us feel sure that the laws of elementary physics would continue to be obeyed, our confidence that the next observed puddle would fail to boil ought now to be severely eroded.

Notice also that Lewis's point about partitioning could lead to severely counterintuitive results in ethics. Imagine an infinite number of mansions each inhabited by three happy people and ninety-seven miserable ones. You learn that some demon created these people. The demon had the power to fill each mansion with people whose distribution of happiness and misery was the reverse: ninety-seven happy people to three miserable ones. None the less, he decided not to. An evil decision, surely—yet our ability to choose different ways of partitioning the people could suggest that it would be wrong to declare that 'the real proportion' of happy people to miserable ones was worse than it might have been. Following Cantor, many a mathematician would reject any such declaration as meaningless. My suggestion, in contrast, is that we need some understanding of the word 'proportion' that would permit it. The demon's decision would be evil, and how else than by making it true that there were, in some crucial sense, proportionately fewer happy people than there might have been? Cantor's way of defining 'equality' in the case of infinite sets may not be the only possible way. Henri Poincaré was firmly against it calling Cantor's approach 'perverse' and 'pathological' (Dauben 1979: 1).

Arguments in this area become very complicated. I cannot claim to have proved beyond all doubt that Lewis is wrong. Still, both his system and my own seem in fairly grave difficulties. My preferred way of escape from them is to say that not all knowledge would be worth having. Detailed knowledge of messy worlds, such as ones which suddenly began behaving crazily, would be knowledge *not* worth having. A divine mind therefore wouldn't contemplate such worlds, and no pantheist need fear that he or she inhabits one of them. However, before investigating this we should look at a curious suggestion. It is that unlimited divine knowledge is not even possible, for reasons to do with the mathematics of the infinite.

Does Unlimited Knowledge Run into Cantorian Difficulties?

According to Patrick Grim, Cantor's treatment of infinities reveals that there cannot be any such thing as *the set of all truths*. From this it follows that divine knowledge would necessarily be limited. There have to be truths, infinitely many truths, that God does not know (Grim 1991).

God's knowledge could still be infinite in a sense not meaning 'unlimited'. On a mathematically standard understanding of what 'infinite' means, a mind's knowledge would be infinite if the mind knew the weight of every single cabbage in an infinite line of cabbages while remaining ignorant of the existence of carrots. 'Infinite' does not have to mean 'including everything'. What Cantor is usually taken to have demonstrated is that there are levels of infinity. Infinite numbers of ever increasing size are reached as we climb to higher and higher levels. What is more, the notion of a highest-numbered infinity is nonsense. So, Grim concludes, Cantor has destroyed the idea of a divine mind whose knowledge is absolutely unrestricted.

Grim's reasoning can be opposed on various grounds, though. For a start, it can be urged that whether one infinity should be treated as greater than another, even at the lowest of the levels recognized by Cantor, ought to depend on what background story we tell. Dead, you are informed by the devil that before entering heaven you must read through an entire library of his. It has infinitely many books. Would it speed things up if you received his permission to read just the odd-numbered ones? Unfortunately not. In this story, the set of *all the books* and the set of *the odd-numbered books* are of identical depressing hugeness. But now, suppose instead that continuing to read books from the devil's library is your sole means of keeping him from carrying you off. Floodwaters are approaching. The odd-numbered books are near the ground and will be rendered unreadable. Why not say that *in a sense*—not the sense of interest to Cantor but a useful sense all the same—the infinity of all the books, including those which will survive the flood, is 'an importantly larger infinity' than the infinity of those near the ground?

Duns Scotus noted that, given two concentric circles, absolutely any point on the larger circle could be 'paired off' with another on the smaller. Take a point at random on the larger circle. Draw a line joining this point to the circle's centre. The point at which the line cuts the smaller circle is the required other half of the pair. Intrigued, over three centuries later, by a similar truth concerning how whole numbers could be paired off with their squares, Galileo concluded that the words 'equal', 'greater', and 'less' were simply not applicable to infinite quantities. But Cantor chose differently from Galileo. When there was any way whatever of pairing off all the members of one set with the members of another, he chose to treat this as immediate grounds for calling the sets 'equal in number'. In other words he defined 'equality in number' in terms of the possibility of finding one-to-one correlations between the members of various sets, even when the sets were infinitely large. He *stipulated* that this was how the phrase was always to be understood. Cantor's definition is often fruitful mathematically: so much so that it has become standard among mathematicians. Yet this does not guarantee that it entirely lacks the kind of arbitrariness that characterizes so many definitions when they go beyond previously well-established usage.

It is, after all, unclear to what extent Cantor's results apply to anything 'in the real world', that is, anything beyond what happens in symbolic systems when the symbols are manipulated according to various rules, the symbols themselves being defined by stipulating which rules apply to them. Writing about this area, I was much too quick to say (1995*a*: 'Finite/Infinite') that mathematicians (*all* mathematicians?) happily accepted that an infinite hotel whose every room was filled could still welcome infinitely many further guests, thanks to the hotelier's ingenious shuttling of guests from room to room. The fact is that even David Hilbert—who admired Cantor's approach and used the hotel in question to illustrate it—said that Cantor's results might have no application to realities outside pure mathematics. When a weary traveller arrives at Hilbert's already filled hotel, it is tempting to think that moving the guest previously in room #1 to room #2, and the one in room #2 to room #3, and so on, would only defer the hotelier's problem. It might defer it so successfully that no symbolic manipulations could ever produce the message that the problem would prove insuperable at such and such a stage, but how could it solve it? The sheer fact that a concept is fruitful mathematically is no guarantee that it corresponds to a reality. (Perhaps *the square root of minus one* illustrates this, but also perhaps not; one sometimes hears, for instance, that its use can correspond to rotating the axes of one's graph. For a less controversial example, consider *negative probabilities*. Physicists occasionally find that bringing these into their calculations speeds them to the right answers[7] yet this could scarcely show that there can

---

[7] Richard Feynman, 'Negative Probabilities', Hiley and Peat 1987: 235–48.

be probabilities lower than zero. 'Zero probability' means utter impossibility, and you cannot have anything less probable than that.)

Possibly more crucial, though, is that any ultimate limits to knowledge which Cantor may have demonstrated *might concern nothing more than truths collected into sets*. It can be argued that Cantor (who was deeply religious) recognized this and concluded that God's knowledge was without any significant limits. The crucial point is that not all collections can be called 'sets'. Here is a standard way of proving it. Some sets are members of themselves: the set of things identifiable in plain English, for example, is itself identifiable in plain English, which is how I have just now identified it. Other sets are not. The set of rabbits is not itself a rabbit. Now, consider those sets that are not members of themselves. Can they be collected together into a set of all sets that are not members of themselves? No, for this would lead to a contradiction. Compare how there cannot be an adult male barber who shaves all and only those adult males of his village who do not shave themselves. Were there such a barber, then there would be no consistent answer to who it was that shaved him.

True, one of Cantor's letters calls the totality of all that is thinkable 'an inconsistent multiplicity'. But this, it can be held, did not mean he believed in limits to what God could know. What it instead meant was that God's thoughts could not form *a complete set*, on the usual technical definition of what *a set* has to be.

I have derived this way of reacting to Grim from work by Plantinga, J. H. Sobel, and A. W. Moore.[8] While it perhaps cannot be claimed that Cantor ever said precisely what I am suggesting, it seems compatible with what he did say and in particular with the following passage:

The actual infinite arises in three contexts: first when it is realized in the most complete form, in a fully independent other-worldly being, *in Deo,* where I

---

[8] Plantinga and Grim, 1993; draft material that Sobel generously let me see; Moore 1990, 1995. Moore repeatedly recognizes his indebtedness to Ludwig Wittgenstein. The latter was highly suspicious of Cantor's alleged discoveries that some infinite sets were really bigger than others, whereas various further infinite sets were despite appearances really of the same size, when these discoveries were interpreted as concerning realities beyond those of *how various symbols were conventionally manipulated*. See Moore 1990: 139–40, for instance, with their quotations from Wittgenstein 1976, 1978.

call it the Absolute Infinite or simply Absolute; *second* when it occurs in the contingent, created world; *third* when the mind grasps it *in abstracto* as a mathematical magnitude, number, or order type. I wish to make a sharp contrast between the Absolute and what I call the Transfinite, that is, the actual infinities of the last two sorts, which are clearly limited, subject to further increase, and thus related to the finite. (Cantor 1932: 378, as tr. at Rucker 1983: 10)

Here, Cantor can be interpreted as telling us that there are some infinities (for example, the infinity of things that a created world might contain) which can be ranked by applying his 'possibility of pairing off' criterion of whether one infinity is equal to or larger than another. These infinities, however, are all of them surpassed by an infinity found in God (*in Deo*), an infinity that is not limited, not 'subject to further increase'. What if we granted that God could never know any reality describable as 'the set of all truths' because (on a standard way of understanding what is meant by *a set*) this would be like knowing a triangular circle or a wifeless bigamist? The sentence 'Every member of the set of all truths is not a falsehood' would then have to be rejected. All the same, we could surely deny *that at least one truth is a falsehood*. And similarly, we might feel inclined to deny *that there is at least one truth which God does not know*.

Unlimited Knowledge Could Well be Undesirable

A pantheist (or even just a believer in a supremely knowledgeable Creator) could reasonably accept that divine knowledge extended not just to immensely many universes, but to vastly many other things as well—to vastly many chess-like games, for instance. Although less deep than *shogi* (Japanese chess) in which captured men can return to the board to fight against their former allies, western chess is a superb game, making it pleasant to think that the divine mind contains knowledge of all possible sequences of moves in it. But how about games rather similar to it? Would it be good to contemplate every last possibility here? In his *Encyclopedia of Chess Variants* David Pritchard (1994) tells us that many thousand such games have actually been developed by humans. They include two mind-benders of V. R. Parton's invention: 'Alice', in which men repeatedly 'pass through the looking glass' between one board and another, and the crushingly complex 'Ecila' in which a six-dimensional board is simulated. An entire journal, *Variant Chess,* is devoted to such possibilities in ever increasing numbers. In a recent issue, for example, Pritchard discussed what he called the variant of the decade. Named 'Hostage Chess', this is a new means of fusing western chess with *shogi,* which it does by allowing exchanges of prisoners.[9] Yet is Pritchard right in his statement that the number of possible chess variants 'is infinite'? To make him right, we shall have to count even variants that are immensely complex, soon reaching ones no human could understand. As is shown by quantum field theory, a lump of matter the size and mass of a human could encode only about a billion trillion trillion trillion *bits,* the simplest possible elements in any message (see Tipler 1994: 407–11; or Moravec 1999: 166). This no doubt limits how far human thoughts could conceivably extend. God's mind, in contrast, might be equal to the task of grasping infinitely many chess-like games of ever increasing complication (one can keep adding more dimensions to the chessboard, for a start) together with absolutely all positions reachable when playing them. Would it really contemplate the lot? Pritchard confines himself to listing about one and a half thousand variants, commenting that anyone can invent another in ten seconds 'and unfortunately some people do'. Could it be good for the divine mind to plough through all the hopelessly unsatisfactory games that humans have come up with, considering all

---

[9] Western rules, except the following. Each player owns two areas at the side of the board—a prison for captured men, near the player's right hand, and an airfield near the left. In each turn you (i) move normally, or else (ii) rescue one man from the enemy prison by transferring one of equal or higher value from your prison to the enemy airfield, then at once parachuting the rescued man onto a vacant square, or else (iii) parachute one man from your airfield. (Values run from *pawn* upwards to *knight or bishop,* then *rook,* then *queen.* Pawns cannot parachute onto first or eighth ranks, but parachuting can place your bishops on squares of the same colour. Pawn jumps from the second rank and acts of castling can involve parachuted men regardless of their earlier positions or movements.) A seventh rank pawn can move forwards or give check only if able to be promoted by changing places with a piece in the enemy prison. See the chapter 'Hostage Chess' in Pritchard 2000.

possible moves in each of them before moving on to infinitely many others which were yet worse?

Again, how about endless sequences of the utterly tedious kind? (i) The word 'that' can start an English sentence, can't it? That *That* can start a sentence is a fact. (ii) That *That that* can start a sentence is therefore another fact, as just now demonstrated. (iii) That *That that that* can start a sentence is therefore yet another fact. . . . And so on, in an infinite series. Well, does the divine mind contemplate every member of the series? And does it next contemplate the point that the member in question is one truth among many, and then the point that it is a truth that it is a truth, and then the further point that it is a truth that it is a truth that it is a truth, and so forth?

In addition to contemplating our universe in all its details, does the divine mind keep track of the distance not just between each particle and its nearest neighbour, but also between it and its next nearest neighbour, etc.? Further, is it vividly conscious of the radius in inches and in millimetres of the smallest sphere that would include each particle and its 773,004,229,924 nearest neighbours, plus—for how could this be avoided if the divine mind knew *absolutely every truth*?—the result of replacing inches or millimetres by a unit defined as 0.136 per cent of the distance between God's outstretched finger and Adam's in Michelangelo's *Creation of Man*?

Is a great deal of God's knowledge to be compared with the dreadful Library of Babel imagined by J. L. Borges, filled with infinitely many books in which all possible arrangements of the alphabet's letters can be found? Hugely many of the books would be intelligible, of course. (An immortal parrot pecking randomly at a typewriter would take only about $10^{3,000,000}$ years—write down one followed by three million zeros and you'll see how brief a period *that* is—to peck out Conan Doyle's *The Hound of the Baskervilles*: Crandall 1997: 77.) But through what oceans of nonsense you would have to trawl before finding an intelligible sentence! Is God in eternal contemplation of every single page?

As well as thinking about utterly meaningless combinations of sounds, does God think about all possible bad radio plays, incompetent performances of awful music, ugly distributions of paint blobs on canvas, unrelievedly boring spells of imprisonment? Does God consider exactly how it feels to be done to death not merely in all the manners in which humans have done one another to death, but in all possible further manners, in the bodies of all possible physical organisms? Besides experiencing all the sorrows that humans have actually endured, does God contemplate (in full detail) every other sorrow in an infinite number of further universes? And does God know precisely how it feels to be intelligent living beings of sorts which could never have evolved in any universe but which are still (unlike triangular circles) logical possibilities and which are in terrible agony, both physical and mental?

While it is hard to be confident of anything here, it can at least seem quite likely that a divine mind *would not* be the better for being conscious of absolutely all facts about actual or possible situations. Is God aware of precisely what would have resulted if at noon yesterday a sunflower had suddenly appeared in your hand, from nowhere? If the third pebble your foot collided with last week had become 4.83 times heavier just before the collision? If the world's largest ruby had suffered some slight change in its colour at that same instant? And does the divine knowledge extend to a world in which you suddenly vanish or turn into syrup? The answer to all such questions could be 'No'.

Unlimited Knowledge Might Perhaps Be Impossible on Grounds Going Beyond Mere Logic

It could well be, then, that my pantheism is not in trouble over inductive reasoning. The divine mind might not keep track, even, 'of exactly how many molecules are discarded when people file their fingernails', which Grace Jantzen (1984: 83) gives as a case of something too trivial to bother with. The matters which God contemplates in detail are those which are worth contemplating in detail. Worlds in which the laws of physics suddenly break down are not. Any knowledge worth calling divine would no doubt include a recognition that immensely many such worlds were possible, but it wouldn't involve knowing their structures in full. Pantheists need therefore have no fear of finding themselves inside one of them. (There could still be some very slight danger of finding oneself in a situation which suddenly behaved in one or other of the bizarre ways that physics cannot rule out, yet this hardly worries me. The number of years needed for quantum fluctuations to topple a beer can on a level surface vastly exceeds the number required by our *Hound of the Baskervilles*-typing parrot.)

While this is my preferred means of maintaining my trust in induction, other means might work as well. Lewis might be right about infinities. Or it might perhaps be that the divine knowledge was limited by factors which were not mathematical (as imagined by Grim) and also not ethical. Perhaps it really would be good for God to contemplate infinitely many worlds in which inductive reasoning failed, if only it were possible to contemplate them *as well as* all those in which it didn't fail, all the beautiful symphonies, all the elegant chess variants, etc. Yet contemplating more than a limited amount may be impossible in point of fact, on grounds whose nature we could at least suggest.

What might the grounds be? Well, it might be impossible to keep compressing more and more complexity into any mind sufficiently unified for its consciousness to form *a whole worth having for its own sake*. The divine knowledge would have immensely many components, but wouldn't they be 'mere abstractions', rather as a stone's shape, its colour and its length are abstractions instead of existing separately? There is nothing too outlandish in this idea. Quantum physics on the one hand, and our experience of our mental states on the other, may give us (see Chapter 2) grounds for denying that the elements of complex situations *are always separate in their existence*. Nevertheless we have little insight into how this would be possible. From which it follows, presumably, that we have no right to be confident that anything, even a divine mind, could be an arena of absolutely unlimited knowledge while remaining adequately unified.

Look again at the sort of thing—it may strike you as ludicrous but it cannot be avoided—that such unlimited knowledge, *knowledge of every single truth*, would have to involve. Consider a mind which, as well as contemplating some particular apple in all its details, was keenly aware of the fact that the apple's mass was 45.364 times that of a particular worm on some distant continent, plus the fact that 45.364 was a number 8.79 times smaller than the length in centimetres of some particular rock on the surface of Venus, and also the truth that expressing all this in Portuguese words and Arabic numerals would require a minimum of such and such a number of characters and that these, on some particular computer screen with the font size set to 12 in the typeface Letter Gothic, would extend exactly such and such a number of inches, a number standing in such and such a ratio to the number of atoms (several billion, believe it or not[10]) which ever became incorporated into the body of Johann Sebastian Bach after once forming part of the horse mounted by Vercingetorix at Alesia, etc., etc. Would it truly be possible for all of this—extending off to infinity in all directions in fantastically bizarre and entangled ways, with each new fact standing in countless new relationships to every other, each of these relationships itself being a new fact standing in countless further relationships— to be crammed into any whole that was unified in its existence? Heaven only knows. Perhaps not even a divine mind could contemplate absolutely every truth because its existence couldn't remain unified while being more and more spread out like a sheet of gold hammered ever thinner.

Pantheistic Writings

'Pantheism' is a word that has been given numerous different senses: for a survey of them, see a recent book by Michael Levine (1994). Still, one can say that what I have been defending certainly deserves the name 'pantheism' and that something at least vaguely like it can be found in many other places—in Hinduism, for example, and particularly in the Upanishads; in much Jewish thought, which greatly influenced Spinoza; in much of the Hegelian thought that started where

---

[10] See Heidmann 1992: 40–1. Of the living matter existing on our planet at a time earlier than a thousand years ago, almost every gram has bequeathed about a thousand hydrogen atoms to your own body of today, there being so immensely many atoms in each gram.

Spinoza left off, although perhaps not always improving on his ideas;[11] and, of course, in Spinoza's own writings.

'Of course'? Unfortunately, hardly anything is agreed nowadays in the field of Spinoza scholarship.[12] You will actually find one of the main authorities suggesting that Spinoza, when he insisted that the universe was a single, fully unified entity, meant only that it was obedient throughout to a single set of laws. My notion that the material world is just God's thinking about physical structures in all their details—which is the interpretation I put on Proposition Seven of Part Two of Spinoza's most famous work, the *Ethics,* that 'the order and connection of ideas is the same as the order and connection of things', with its Scholium commenting that this had been 'glimpsed by those Hebrews who hold that God, God's understanding and the things which God understands are all one and the same thing'—will be sure to strike some of the authorities as in conflict with Spinoza's actual position because of giving too much primacy to God's attribute of mentality. There are those, too, who argue that by 'God' Spinoza simply meant the universe as a unified whole, while by 'goodness' he intended something like 'degree of completeness of being', period. To all which I can only reply that, despite all the definitions, axioms, propositions, demonstrations, corollaries, lemmas and so forth deployed in his *Ethics,* and also despite his being a hero of mine, Spinoza seems to me rather an unclear thinker. If you want to see pantheism defended clearly, it

---

[11] The kind of Hegelianism once popular in Britain may be as interesting as anything Hegel himself wrote. F. H. Bradley's *Appearance and Reality* is particularly famous but A.E.Taylor's *Introduction to Metaphysics* is much easier to understand. Let me emphasize, though, that I dislike many typically Hegelian claims: for instance, that reality is often utterly different from how it appears to be, or that events are never tragic in any ultimately important way.

[12] Major disagreements on interpreting Spinoza centre hardly at all on how individual sentences should be translated—something which can also be said about the other philosophers whose translated words appear in this book. With a large variety of published translations to choose from, the book still sometimes gives its own wordings, taking full responsibility for these as well as for the accuracy of any suggested by other people instead of repeatedly sending the reader to numbered notes. However, in the case of Spinoza those wanting to read more could well go to Curley 1985. Other useful sources include Shirley 1982 and Wolf 1910. And Victorian renditions retain their charm. Slightly revised, they can be found in Gutman 1949.

would be far better to look at Sprigge's masterpiece, *The Vindication of Absolute Idealism*.[13] I claim only that the sort of thing I defend might reasonably be called 'Spinozistic'.

As an illustration of how hard it is to interpret Spinoza, consider whether he thought that God's reality included absolutely all possible things. It is very commonly said that he did think this. People point to Proposition Sixteen of Part One of the *Ethics*. This tells us that from the necessity of the divine nature 'infinitely many things must follow: that is to say, all the things which can be conceived by infinite intellect'. And yet, look next at Proposition Eight of Part Two. Here we are told that the ideas of *non-existent* individual things are included in the infinite idea of God. Well, how can anything at all be non-existent if infinitely many things are conceived by the divine intellect, all of them therefore 'following' (which must surely be taken to mean that they don't remain merely possible)? Must we understand Spinoza as saying that, Yes, God does think of infinitely many things, but No, he doesn't think of all possible things? That the situation is as if God thought of infinitely many cabbages but never of carrots? Well, in a Note to this Proposition Eight we are offered an analogy. Spinoza asks us to consider a circle which can be thought of as including infinitely many rectangles formed where chords intersect, rectangles which have interesting properties. Of the intersecting chords we are asked to conceive two only as *existing*. This could seem to show that he distinguishes between two ways in which things could be included in God: being included as possibilities that God understands—genuine possibilities with definite properties—and being included as truly existing things. But how could that be made consistent with Proposition Sixteen of Part One?

I must leave the point to the experts—if indeed anyone, even Spinoza himself, could ever have been expert about any such thing as 'how Spinoza's system really runs'. It has been known, after all, for truly great philosophers to defend contradictory positions when the arguments tug first in one direction, then in another.

---

[13] Sprigge 1983. See also Sprigge 1984: ch. 8 ('Spinozistic Pantheism'), and Sprigge 1997. Sprigge has written to me that, in the sense in which I am a Spinozist, he is one too.

Minds Human, 2
Artificial, and Divine

Although carrying an infinitely complex pattern of thought, a divine mind could be unified in its existence. Its individual elements would in that case be abstractions, somewhat as a tree's height or a flower's colour are abstractions (while still being fully real).

All the same, such a mind could contain many elements which looked very much as if each had an existence truly separate from that of the others. Some divine thoughts would give the structure of material objects in complete detail. As explained earlier, it could then be argued that they actually were such objects—yet this wouldn't make all material objects so closely unified with one another that they all seemed mashed together. Similarly, divine thoughts about individual people could actually be those people without everyone knowing what everybody else was thinking.

(1) The chapter looks closely at how the divine knowledge could include 'areas filled with ignorance' that were the thoughts of individual people, and at how there might be a divine overview in which the contents of those areas could be known 'as if telepathically'—the idea being that you might (if telepathy actually worked) be telepathically aware of somebody's ignorance without yourself being ignorant, although getting really quite a good idea of how it felt to be ignorant. (2) In addition the chapter examines unity of the kind known to us through introspection, suggesting that it is essential to the value of our conscious states. Quantum physics might

throw light on it and also on the unity of 'quantum wholes' in general, wholes which might sometimes be very large. But this is far from the same as saying that everything—even a divine mind—that possessed such unity would have to obey all the laws of quantum physics. (3) Even if rejecting pantheism, might not a philosopher view all things as needing to be 'made out of mental stuff' in order to exist at all? Perhaps nothing can exist unless it has some degree of complexity, the very simplest of things still consisting of various elements unified in a thoroughgoing way that is worth calling 'mental' because consciousness of some primitive sort is necessarily involved. (4) However, it could well be a mistake to argue that entities related to one another in any fashion, complex or otherwise, must always be mere aspects of one and the same thing. A divine mind of which we were parts might be one among infinitely many.

Be warned: the topics of this chapter are complicated. Descartes apparently thought that the nature of consciousness was obvious to every conscious being. Now, while that could be correct in one respect, in others it is clearly wrong.

Structural Unity and Unity of Existence

The divine mind of which you and I are supposedly parts was pictured in the previous chapter as supreme in its value. The divine thoughts form a whole supremely worth experiencing. Now, one of the main themes of this present chapter is that any states of mind intrinsically worth experiencing must be more than just collections of elements each with an existence truly separate from that of the others, and each in itself worthless. The basic constituents of such states of mind must instead be abstractions, rather as are the redness and the length of a brick.

We must tread carefully here. First, although in a sense 'merely abstract' the brick's redness and its length are certainly *real*. Furthermore, their reality isn't like that of an abstraction such as the number two or the fact that five threes make fifteen. They are abstract in that it is the red brick, of such and such a length, which is the fundamental reality, while the redness and the length manage to exist only as elements or aspects of that reality. Yes, the redness and the length are distinguishable from each other; yes, each certainly exists; but they do not 'exist separately from each other' in any interesting sense. Second, any collection of things each genuinely separate in its existence would of course be real as well. What is more, it wouldn't be an abstraction in the way in which the number two and a brick's length are abstractions. Still, it would be abstract in another fashion. Suppose a truck is filled with bricks, each with an existence truly distinct from that of the others. The reason why the truckful of bricks exists is then simply that the individual bricks exist in the truck. The existence of the truckful is 'derivative' somewhat as the similarity in length of two of the bricks can be derivative. The similarity exists, sure enough, but its existence is completely secondary to that of the individual bricks. Anybody wanting to create two similar bricks would not need to create first the bricks and then their similarity.

While much of this might appear obvious, the area is a minefield of philosophical difficulties. For one thing, it could be a mistake to think that bricks truly exist separately instead of as aspects of a greater whole. If a Spinozistic, pantheistic approach is on the right lines, then bricks don't in fact exist separately. Bricks are patterns inside the divine thinking, which isn't split up between many separately existing thinkers. The divine mind is a whole with a unity unlike that of any collection of things each separate in its existence—things 'unified' just through, say, being collected together in a truck. Yet the divine thinking does include many elements (bricks, for instance) which at first seem to exist separately. Moreover these are collected into many distinguishable wholes, each of which has *structural unity,* the kind of unity into which things enter when they are organized in various ways.

The central point is that structural unity must be distinguished firmly from *unity of existence*. The unity of a well-trained, well-led army, or even of a collection of ill-trained, ill-led, mutinous soldiers, is one affair; the unity of elements that are all of them features of one and the same existent is another; and this is so whether or not the two kinds of unity are had by one and the same thing. Suppose that an army truly were a pattern of divine thinking. Its soldiers would then be mere aspects of the divine existence, but they would be further unified, sometimes impressively and sometimes not, by entering into such systems as (*a*) the army, (*b*) the human race, (*c*) all terrestrial organisms, (*d*) the things on the surface of planet Earth, and (*e*) the Milky Way galaxy. Labelling all of these 'systems' and saying that their parts have 'structural unity', I mean nothing dramatic. A heap of sand is 'a system of sand grains' in my terminology. The seventy-six pebbles nearest to your left heel form 'a system', although presumably not one of any great interest. Just any collection of things, including one made up of the world's largest statue plus its smallest spider plus your nose and eyebrows, can count as 'a system' in my sense which is deliberately very loose. The most interesting systems, the ones which will tend to inspire us to use printer's ink in discussing 'structural unity', are no doubt those that are (like the well-trained, well-led army) highly organized, but the difficulties of defining 'degree of organization' are not ones in which I care to get entangled. On the other hand, whether things are united in their existence is a question intended to be technically precise. *Things united in their existence* are in themselves abstractions rather as a brick's length is an abstraction.

Without Unity of Existence there can be No Real Value

My position is that structural unity can never give intrinsic value to a state of mind—and states of mind (of a divine being or of lesser beings like us) are, so far as I can see, the only things that could reasonably be thought to have such value. Value is sometimes 'instrumental' rather than intrinsic. Instrumental value might be had by a good television set, a useful instrument for bringing about interesting, entertaining, intrinsically worthwhile mental states. But if it existed without any living being to view it, for example after the annihilation of all life on earth, a television set could be neither good nor bad in any ethically important sense. The idea of a television set whose internal states were of benefit to the television set itself, just through being what they were, is nonsense. In contrast your mental states can sometimes benefit you just through being what they are. Their value really can be *intrinsic*. They can be worthwhile in a respect not dependent on their serving any further purpose (although of course they might do that as well).

The account of intrinsic value I developed in *Value and Existence* and elsewhere[1] owes much to Plato and to G. E. Moore. It includes these elements:

1. Calling things 'intrinsically good' or 'intrinsically bad', I am not just reacting to them emotionally or prescribing that people (myself included) are to favour them or avoid them. Instead I am trying to describe realities of goodness and badness.

2. A thing's intrinsic goodness isn't a quality added to its other qualities like a coat of paint. It is instead a status the thing has: the status of *having an existence that is ethically required to some extent*. This doesn't mean that the thing ought to be favoured in all circumstances. There might be alternatives which had greater intrinsic goodness, or circumstances could be such that the thing's existence couldn't be had without the existence of various other things which ought to be avoided. All the same, it would be an absolute fact that this particular thing would be better than a blank if it could exist all alone. And if it *could not* exist all alone, not even 'in theory', perhaps because a Spinozistic, pantheistic world-picture was right, then it would at least be an absolute fact that something very much like it, something which *could*, would be better than a blank.

3. Talk about such absolute facts in no way implies that human likes and dislikes are unimportant. It is fully compatible with believing that only mental states that include pleasure have any intrinsic worth. None the less, saying they had intrinsic worth would be different from simply saying they included pleasure, or pleasure in company with various other specified ingredients. It would be saying that as an absolute truth—a truth not relative to any particular group's standards because ethics is not just glorified etiquette—these mental states had an existence which was (to some extent) ethically required. Compare how it is absolutely true, a fact of what is mathematically required, that two and three make five.

[1] See Leslie 1979; also, e.g. Leslie 1972, 1996a: 155–70.

4. Admittedly ethical requirements aren't provable by appeal to definitions, like the fact that three plus two must equal five. They may well not be provable at all, in any strong sense of 'provable'. Their existence is assumed by ordinary thought and ordinary language, however. Also, if you believed various bizarre things about them—for instance, that the states of mind most ethically required are those of torturers who enjoy their torturing—then you would belong in a mental hospital. You would belong there just as much as anybody who couldn't grasp that it was required in another way, 'inductively required', to think that very hot water would hurt fingers tomorrow just as it did yesterday.

5. If all realities of intrinsic value were absent from the world, would it then be intrinsically worthwhile to believe that they were absent? Self-evidently not—from which it follows that there'd be no real point in believing it, in any sense of 'real point' that was truly worth anything. What if all belief in intrinsic value were primitive, superstitious nonsense? We could still have enthusiasms and we could be very enthusiastic about our enthusiasms, but this wouldn't make those particular enthusiasms really better—in the traditional sense of 'really better' when the words are used with ethical seriousness instead of in such statements as 'Boiling oil is really better than very hot water for inflicting pain'—than any other enthusiasms or than the state (if that's possible) of having no enthusiasms at all. There would be nothing absolutely wrong in having the enthusiasms, and nothing absolutely right either. Many philosophers of today seem to think this would be a satisfactory situation. It strikes me as a profoundly depressing one. I recognize that, were it our actual situation, then nothing intrinsically worthwhile could be obtained through being depressed by it. But belief in realities of intrinsic value is at the heart of my own eagerness to continue living.

Now, I think my states of mind would have no intrinsic value if they were made up of elements each genuinely separate in its existence. So if they are (as I believe) simply states of my brain, then I very much hope that they are not mere systems of separately existing active brain cells. Still, why is that? Why couldn't collections of separately existing parts come to have intrinsic value *through the structural unity into which the parts fell?* When, as in the case of active brain cells, this unity was complex enough to lead to successful information-processing, why would this be insufficient? Why think some further unity would be required?

In *Value and Existence* I answered (1979: 173) that any assemblage of separately existing things

would be as little unified *in its actual being,* and so for my purposes when I ask whether we have here anything *whose being is ethically required for its own sake,* as any whole made up of fifty-five separately existing blue-eyed left-handed teenagers. True, a whole can have intrinsic value when its parts have it. If each of the fifty-five blue-eyed left-handers has a conscious life with intrinsic value, then the whole set of these conscious lives also has it. But no whole made up of separately existing parts could have it unless some of these parts had it first.

Yet perhaps this argument of mine didn't really prove its point. Might it not be countered that a group of separately existing parts could, by virtue of having the right kind of organization, for example the kind of organization found among the active cells of conscious human brains, form a whole very importantly different from any set of parts that lacked such organization? Why, then, dismiss it as 'too much of an abstraction', 'too merely derivative when intrinsic value is in question'? After all, an opponent of my position could comment, the existence of the whole wouldn't be a fiction. Just like the parts, the whole would genuinely exist. And its existence would be different, genuinely different, from what it would have been if the parts had been differently organized.

It might be impossible to refute this opponent. But it still strikes me that derivative existence—existence of a kind which wholes could have simply through having separately existing parts that were organized in particular ways—is far too abstract for present purposes. The point may not be provable firmly, but few philosophical points of much importance are provable firmly! Imagine that some demon examines the cells in a conscious human brain, then replicates their activity-patterns with the help of other brain cells (let's call them *imitating neurons*) each of which is *firmly separate in its being.* If need be, the demon would ensure the firm separation by constructing each new one of the imitating neurons from material particles in a new universe with a space all of its own, which is a notion few modern cosmologists would find baffling.[2] The idea is that if a particular cell from the human brain in question had, during a particular brief period, successive moments of activity and inactivity—of 'firing' and of failing to fire—in, say, the sequence

1000101001110010011000001010011100101011100010110011100001110100111101

then the demon would ensure that one of the imitating neurons also had periods of activity and inactivity in this same sequence. What's more, he would ensure that all the periods of activity and inactivity of the brain cell, over perhaps several minutes, were similarly replicated by its imitating neuron, and that those of absolutely all the other cells in the brain in question were replicated as well. Hugely many imitating neurons would be used. Now, while almost everybody would agree that the replications would have no intrinsic worth, quite probably *not even this* would be solidly provable. After all, an opponent could comment, the entire system of replicated brain-cell firings wouldn't be a fiction; its existence would be genuinely different from what it would have been, had its parts been differently organized; etc. But while such comments might not be provably wrong they could well be thought absurd.

Quantum Theory can Suggest that our Universe is Unified in its Existence

By 'a single existent' let us mean something that is more than a collection of things each with an existence genuinely separate from that of the others. Another label fitting it would be 'an existentially unified

---

[2] When first running the argument—in 1970 in the *American Philosophical Quarterly*—I unfortunately watered down the main point by leaving out the words 'firmly separate in its being'. I wrote simply of intelligent beings *scattered through our own universe*, each controlling 'the discharges of a human brain cell transported to his planet'. Later variants developed by other philosophers include scenarios in which cerebral activity is simulated with the help of vast numbers of people each carrying out some very simple task: e.g. see Block 1978.

whole'. In our terminology, a single existent will be something such that any components that it has are abstractions rather like the length of a pond or a ripple on its surface, or the grin on a face. Now, why (unless we are already pantheists) should we think that all the things in our universe are simply elements or aspects of a single existent?

We could answer for a start that we were impressed by what physicists say. David Bohm writes that 'on the basis of modern physics even inanimate matter cannot be fully understood in terms of Descartes's notion that it is nothing but a substance occupying space and constituted of separate objects'; 'the world cannot be analysed into independent and separately existent parts'; the interaction of particles in a many-particle system can often best be thought of 'as depending on a common pool of information belonging to the system as a whole' (1990: 272, 275, 280). His evidence for this lies in well-established phenomena in quantum physics, such as the following.

1. At the low temperatures at which superfluidity and superconductivity occur, particles move together in patterns which cannot be disrupted by the kinds of tiny obstacle that usually cause friction and electrical resistance. The obstacles might be expected to disrupt the movements of individual particles. Now, while particles cooled to superconducting and superfluid states don't entirely say goodbye to their individuality, since it makes sense to ask how many of them there are in particular regions, they still (as Bohm emphasizes) say goodbye to enough of it for resistance and friction to be abolished.

2. Superconducting quantum interference devices, SQUIDs, are superconducting rings containing narrow constrictions. Although they can be as large as thumbnails, SQUIDs have been described as behaving 'as single quantum objects'. The electrons in a SQUID can occupy the same quantum state, which permits huge numbers of them to quantum-tunnel together across a constriction: there can be readily detectable changes in a magnetic flux produced by something like a hundred billion trillion electrons, changes occurring with the spontaneity characteristic of quantum mechanics.[3] Phenomena such as

---

[3] Spiller and Clark 1986; for more details see Clark, 'Macroscopic Quantum Objects', at Hiley and Peat 1987: 121–50. Shimony 1988: 52–3.

this follow from the fact that the particles involved have wavelike characteristics, and that at low temperatures it becomes easier for their quantum wave functions to overlap. To the extent to which they do overlap, the particles 'become indistinguishable'; there can be 'long-range order and a sharing of the identity of constituent units' in states that in extreme cases become describable as 'Bose–Einstein condensates' in honour of the two scientists whose work led physicists to expect their existence (Marshall 1989: 78–9). In 1995, thousands of atoms cooled to within about twenty billionths of a degree of absolute zero became what was dubbed 'a single superatom' for several seconds. Ever larger and longer lasting condensations have been produced in subsequent experiments.

3. Some particles (photons, for example) are bosons and obey 'Bose-Einstein statistics'. What's the probability that two bosons in the same quantum state will be found in different halves of a box? If the particles could be treated as each fully separate in its existence, then the answer would be ONE-HALF, the four possibilities being (*a*) *particle P to the left and particle Q to the right;* (*b*) *particle Q to the left and particle P to the right;* (*c*) *both particles to the left;* and (*d*) *both to the right.* Experiments indicate, however, that the first two cases are identical—identical in some far stronger sense than just that humans cannot distinguish them—because the actual probability that the particles will be found in different halves is ONE-THIRD. *P's being in the left half of the box while Q is in the right half* seems not to be genuinely different from *Q's being in the left half while P is in the right half.* To calculate the probabilities correctly, we must treat these seemingly distinct alternatives as one and the same. It is tempting to describe each particle as 'bi-located'.

4. In experiments inspired by J. S. Bell, pairs of particles originating at a single point, perhaps when another particle decayed, turn out to have properties which can remain strongly correlated no matter how far apart the particles move. When one particle in such a pair is an electron and is found to be 'spin up', then its distant partner will—if it has not previously interacted with anything—be found to be 'spin down', no matter which direction is chosen in defining 'up-ness' and 'down-ness'. (Taking any straight line you

like, specify that it shall be the axis of up-ness and down-ness. If the one particle is found to be spin up relative to this freely selected axis, then the other will be found to be spin down, always.) And something similar applies in the case of paired photons. If one of them hits a polarizer oriented in an arbitrarily chosen direction and gets through it, then the other, no matter how far it has travelled, will—if it hasn't previously interacted with anything—get through a similarly oriented polarizer without fail. This is so despite the fact that photons that are known to be all of them polarized in the same direction, because they have been passed through a polarizer to ensure this, react in totally unpredictable ways when they hit a second polarizer angled at 45° to that direction: just half of them get through and the other half are blocked. Theorists cannot readily make sense of these findings by supposing that the particles in each pair have definite spins and polarizations before any measurement. In the case of the spins, the situation isn't like finding the head of a fish in one package, which guarantees that another package must contain not the head but the tail. Before either of them has passed through a measuring apparatus (or otherwise interacted with something, for it would be sheer magic if a particle could know which interactions happened to be measurements made by humans or other intelligent beings) the particles simply do not have spins and polarizations individually; their properties are 'entangled'. Now, this could seem firm evidence that *these parts*, at least, of our universe *are not separate in their existence.* Nor should they then be judged untypical of the universe. As Paul Davies says, whenever two microscopic particles interact and then diverge 'they can no longer be considered as independently real things' (1980: 125). Lee Smolin correctly insists that 'given any one electron, its properties are entangled with those of every particle it has interacted with, from the moment of its creation' (1997: 252), for once two electrons have entered into entanglement, their later interactions with further particles could only bring added complication to the entanglement. Michael Redhead concludes that in such cases 'the possibility of analysing or reducing wholes in terms of their parts' is ruled out; we are forced to deny 'that the component systems possess their own local properties independently of the holistic context'.[4]

5. We apparently cannot account for entanglements by imagining messages passing at tremendous speed, messages saying, for instance, that because one of two paired particles has become 'spin up' or 'polarized in this direction' with respect to a particular axis, the other must henceforward behave appropriately with respect to that same axis. Experiments reveal that when the axes are chosen just before those particles hit any measuring apparatuses the behavioural correlations survive regardless of how far apart the particles have moved. The matter has now been tested over distances of many kilometres. Not even messages moving at the speed of light could explain such correlations.[5] This, though, shouldn't be viewed as a particularly startling modern discovery for it is only a variant on a phenomenon that has long been recognized. Quantum theorists have long insisted that any firm distinction between particles and waves is an illusion. Particles are sequences of events connected by 'waves of probability'. There can be small regions in which a given particle is almost certain to make its next appearance yet it always retains at least some slender chance of appearing in a very different region: in some immensely distant galaxy, perhaps. (Even if taken just by itself, this might provide strong encouragement for doubting that particles are truly separate in their existence from one another, for how could they be when there is a sense in which each appears to fill the universe?) Still, just as soon as a particle is found in one place it is certain at all other places that the

---

[4] Redhead 1995: 51. Compare Zohar 1996: 443: 'Where classical atoms can only bump, clash, and go their separate ways after meeting, quantum entities overlap and become entangled' because they possess 'both particle aspects and wave aspects', and 'when the extended wave aspects of two similar quantum systems become entangled, those systems begin to share an identity'.

[5] Suppose atoms A and B have spin correlations of the kind described. 'With regard to atom A', Bohm and Hiley write (1993: 138–9), one can say that its interaction with an apparatus set to measure spins in a particular direction 'was "responsible" for the "collapse" on to the eigenfunction of its spin' in that direction; but if there were no connection between the two atoms then how on earth could atom B 'know' that it too must collapse into an opposite eigenfunction of the spin in the same direction, even though this direction 'was chosen arbitrarily when someone decided to measure A when the atoms were far apart', the choice being made 'too rapidly for a signal to pass between A and B'?

particle cannot be found there instead. This is a fact with which physicists feel fairly comfortable, granted that—as is almost universally accepted—no faster-than-light *signals* could ever be sent by exploiting it, so that relativity theory isn't violated. It could well be a sufficient excuse for talk of faster-than-light *influences,* however, since it does at least show that far separated regions can be connected much more intimately than common sense would suspect.

6.   Naturally enough, phenomena of the kinds just now mentioned are most evident in the cases of those particles that are most clearly associated with waves: for instance, the photons of visible light. Sir Arthur Eddington asks us to consider 'the light waves which are the result of a single emission of a single atom on the star Sirius'. By the time the wave front reaches our planet, the energy of those waves 'would seem to be dissipated beyond recovery over a sphere of 50 billion miles' radius'. Suppose, though, that the waves come to affect somebody's eye. The entire energy in the wave front must then concentrate itself into that eye. It must become certain everywhere else that the waves cannot be detected there instead, despite there being nothing that moves at faster-than-light speed to all regions of the wave front carrying the message (in Eddington's words) 'We have found an eye. Let's all crowd into it.' Yet, Eddington notes, 'well-established phenomena such as interference and diffraction' appear to rule out the possibility that at the start of their journey the light waves somehow themselves knew the route they were going to take, words about a spherically expanding wave front being therefore just a colourful way of describing human ignorance. It appears that we must instead accept that the light waves really do 'carry over their whole front a uniform chance of doing work' (1928: 185–9).

Common Sense Itself Indicates that Something Unified in its Existence can be Complex

Physicists sometimes argue as well that particles are best pictured as tiny wrinkles or ripples of space or of space–time, which would make them as abstract as any larger wrinkles or ripples. But leaving physics aside and appealing simply to common sense, we could still find plenty to support the idea that a whole unified in its existence—one whose components were all of them 'mere abstractions' rather as are a length, a ripple, or a grin—could none the less manage to be fairly complicated. There are actually arguments suggesting that unless a thing had some degree of complexity it could never exist at all *except as* a mere abstraction, an aspect of something else. Here are some commonsensical points we could make:

1. Anything possessing a spatial position must obviously possess something else in addition, in order to make the distinction between its being at a particular place and there being nothing at that place; so here, for a start, is some minimal complexity.

2. People often maintain that nothing could be positioned in space unless other things, too, were positioned in it, so that it could be near to them or distant from them. It is then only a short step to the idea that all spatially positioned objects must be linked at least partially in their actual being. Each, we might say, couldn't be the kind of existent it was unless it possessed a spatial position; however, it couldn't possess one if the others were annihilated; hence, we could be tempted to conclude, it would be too abstract to exist in absolute isolation. If we tried to resist such a conclusion by imagining an object surrounded by nothing but empty space but still 'positioned at the place where it was', then that would presumably involve thinking of the empty space as itself an existing something (for how else could it do any surrounding of anything?), in which case the conclusion would follow just as before. To have a hope of blocking the conclusion, we should have to imagine the space in which the object extended as a space coming to an end precisely at the object's boundaries, which could seem far too odd.

3. Besides, what kinds of entity could we have in mind in talking of 'entities that are entirely simple'? Would we be trying to picture *point-events* having absolutely no spread either in space or in time? It could well be judged that nothing quite this simple could possibly exist. How could anything with an existence spreading over no period whatever, something lasting for no time at all, be any different from nothing?

Again, if an entity had no spread in physical space or in something else worth calling space—perhaps an experiential or perceptual space in which an afterimage produced by a bright light could be said to be 'a round afterimage' or one which was 'very close to' or 'smaller than' another afterimage—then could it be any more real than something which existed for no period? I suggest that it could not. Provided we were suitably open-minded about what might count as 'extension', we could well think Spinoza correct in the assumption Peter Loptson finds running throughout his writings, that 'unextended reality is unthinkable, logically impossible' (1988: 32). (In his *On the Improvement of the Understanding* Spinoza calls it a grave error to believe 'that the parts of extension are really distinct from one another'. Being extended certainly doesn't strike him as being a mere collection of separately existing point-entities.)

4. In any case, we might want to ask, how could any point-events, or any particles extending in time but not in space, ever get to know about one another's existence? How could they so much as learn they were in one and the same universe? If the existence of each were totally separate from that of the others, what would tell each that the others were there to be interacted with?

The right inference, I suggest, is that the world as we know it, and probably also absolutely any world that is more than a mathematician's invention, must be composed of one or more objects, each somehow *spread out in its very existence* so that it doesn't owe its extension to being a mere group of entities which are each of them entirely extensionless. But if so, then this raises an interesting question. Why think that any such object *would be forced to have exactly the same characteristics at all the points over which its existence extended?* Why, that's to say, should the mere fact that the same single existent (defined as something having no components other than mere abstractions) *occupied all the points* be thought to imply *that every point took on precisely the same features?* Why cannot a single existent have complexity of the kind had by our universe, a complexity not just of many qualities, but of qualities that differ from place to place? Here we enter territory fought over by Bertrand Russell and F. H. Bradley.

Bradley, Russell, and Relations

How, Bradley asked, could two things come to be related (by, for instance, being 7 feet apart) if they were two separate existents? His view was that all relationships needed to be treated like the one linking 'the greenness and the other perceived qualities' in the case of a green leaf. The greenness was a mere element in a unified reality, the leaf, and the unity of that reality wasn't an affair of mere 'relational coupling' such as people might fancy to be all that united the grains of a sand pile. Instead, the plain fact was that any leaf's colour, its shape, its volume, and its flexibility, were abstractions incapable of existing independently. And the same was true, he claimed, of absolutely any elements of our immensely complicated universe. The only thing 'real independently' was a universal whole, the Absolute, a 'super-relational unity of the One and Many', a 'union of sameness and diversity'.[6]

Russell protested that while we might try to explain the relationship of, say, *nearness* that linked thing A and thing B, by claiming that a unified existent, A–B, *contained or was qualified by propinquity,* nothing similar could be done with the case of A's being *larger than* B. Declaring that the unified existent A–B *contained or was qualified by diversity of magnitude* would, you see, leave it completely unclear whether it was A that was larger than B, or B that was larger than A (Russell 1903: 215; 1914: 58). And the same could be said of absolutely all 'asymmetrical' relationships: ones like *being the father of,* or *being a more open-minded philosopher than,* or *being able to defeat consistently at snooker,* where the fact that thing number one is related in such and such a fashion to thing number two means that thing number two definitely cannot be related in this same fashion to thing number one.

Even, though, if Bradley failed to prove his main thesis—even if he failed to show that absolutely all relationships involve unification of the kind which he saw in the case of a leaf and its greenness—it would seem that Russell's attempt to demolish the thesis by pointing to

---

[6] Quotations all from Bradley 1935: ch. 13—an essay, 'Relations', on which he was still at work when he died after years of struggling with the topic.

asymmetrical relationships was just as much of a failure. Consider the colour purple as it appears in a purple amethyst perhaps, or perhaps in a purple afterimage of a bright light (bearing in mind that the afterimage would be purple 'in experiential or phenomenal colour' rather than purple in the sense of being able to interact with light waves in particular ways). When he held that qualities such as being purple were elements of our universe, which was a whole unified in its existence, Bradley meant nothing so silly as that the universe was purple all over. He wasn't intent on denying the obvious truth that the world contains many things and that these differ in their qualities. Still, what if we restrict our attention just to an amethyst or to an afterimage which does happen to be purple all over? The purpleness of this object might be described as a case of 'diversity despite unity' in that it was a mixture of the elements reddishness and bluishness. Now, suppose that every bit of the object displayed precisely the same mixture of those elements. We could still have *an asymmetrical relationship* here. The blue in the purple could be *more dominant than* the red. In many a red-blue object, the red is a good deal weaker than the blue, isn't it?

True, this doesn't prove that an entire world could possess the kind of diversity-in-unity that interested Bradley. The diversity between the red and blue 'constituents' of something purple could strike us as very different from the diversity between a dog and a cat, two ordinary parts of our surroundings. Perhaps anything whose elements are all of them mere aspects or abstractions does have to be 'the same all over'. What we can none the less say is that Russell's argument against Bradley failed to prove it. And direct experience can join with quantum physics in suggesting that regions of one's consciousness at particular moments are wholes which do have Bradleian diversity-in-unity, whether or not the universe in its entirety is itself another such whole of which these regions are mere aspects.

Before tackling this, though, it might be interesting to ask whether a whole's unity-of-existence must be understood quite as Bradley maintained. The parts of the whole wouldn't in fact exist independently; but might other things precisely like them do so, at least in principle?

In a Whole Unified in its Existence, Must a Change in Any Part Affect All the Others?

It seems clear enough that the length and the shape of an amethyst are too abstract to exist separately, and also that an amethyst couldn't continue to exist if it entirely lost length or shape. Still, how about a red-blue amethyst which said farewell to some of its redness? Whereas 'disembodied lengths' or 'disembodied shapes'—lengths or shapes which aren't the lengths or the shapes *of anything*—are utter absurdities, an amethyst deprived of some of its redness could surely continue to exist. And might we not feel inclined to say that it was in all its remaining features *precisely what it had been*, much as we might want to say that a sand pile from which a grain of sand had been taken could be precisely what it had been with respect to all its other grains?

Suppose physicists told us this was impossible ('Even just one sand grain affects all the others gravitationally'; 'we physicists know enough about colour to say that an amethyst cannot change in colour all alone'). Might we not still be tempted to hold that a purple afterimage could lose some of its redness while remaining in all other respects unaltered ('The blue element in the purple could be entirely unchanged')? Or again, might we not suggest that any further alterations which spread through a sand pile, an amethyst, or an afterimage whenever one of its elements changed *should be seen as spreading causally*—like clashes and clangs moving along a line of railway carriages—*owing to physical laws which just happened to rule the world*, instead of with a necessity that was absolute? Might we not want to reject Bradley's dictum (1935: 664) that whenever elements form a whole unified in its existence, every change will necessarily and immediately 'affect the whole throughout and not leave that anywhere unaltered'? Even if viewing our entire universe as a whole whose parts are all mere aspects or abstractions, might we not dismiss J. M. E. McTaggart's claim that the fall of a sandcastle on the English coast would therefore alter the Great Pyramid at least very slightly, the change being something utterly inevitable?

I am unsure how to react to all this. The idea of wholes containing elements that are mere abstractions seems plain enough: a thing's size or its colour are obvious instances. It is also clear that, as a matter not just of causal inevitability but of something stronger, *some* such elements simply cannot be taken away from their wholes while leaving all the other elements unaltered. (An elephant which suddenly came to be without any size whatever wouldn't be merely miraculous, like one which had suddenly grown as small as a beetle or had lost all its colour, becoming perfectly transparent. It would be something non-sensical.) But in the case of anything unified in its existence, would *absolutely any* change necessarily reverberate throughout the whole? Afterwards, all the other elements in the whole would of course be elements *in a different whole,* but would they therefore have to be different in themselves?

The answer is that they would, I very much suspect. Although willing to entertain the idea that the bluishness of a purple afterimage could remain unaltered if the reddishness of the afterimage changed, this does strike me as wrong. 'Bluishness when mixed with a lot of reddishness isn't quite the same bluishness as bluishness mixed with a little', I want to protest. But I cannot see how to settle such matters firmly.

Imagine a tiny dust speck in interstellar space. Suppose that Spinoza and Bradley are right: this speck is merely an abstraction from a unified reality, our universe. It follows that *the rest of the universe* is itself *also* merely an abstraction from this unified reality. Well, does this mean that it would all of it be forced to change at least slightly, were the speck annihilated? Bradley thought so, yet I cannot prove it. While an abstraction *is not* something that exists independently, it could still perhaps be something such as *might have* existed independently. The speck could therefore perhaps vanish and leave the rest of the universe entirely unchanged, at least in principle. (Not in practice! Physicists would be quick to point out that the effects of a single grain of dust aren't nearly as unimportant as people tend to think. As Redhead reminds us, just move a mass of one gram through a distance of one centimetre on the star Sirius and you 'could totally change the molecular dynamics in a sample of gas on the Earth' (1995: 32). That's how sensitive our universe is to slight changes! When you want to know precisely which molecule will be where, the gravitational pull of an immensely distant speck cannot simply be dismissed. But we could accept this and still think Bradley's arguments failed.)

Our Conscious States Show us that Wholes can be Unified in their Existence

That elements in a conscious state can be united to one another in their existence is what many philosophers have believed. Comparatively few join Spinoza in the further step of theorizing that they would be united in this way because they were themselves only parts of a greater unified whole which is our entire universe. It is more typically thought that any unification extends, at most, to all the successive conscious states of a particular person over a lifetime, and perhaps only to all the constituents of any one conscious state at a particular moment. In fact, quite a few of the philosophers doubt whether it extends even that far. Consider everything of which you might reasonably be said to be conscious at this very instant. Should all of it be described as obviously unified in its existence? 'Very possibly not' would be their answer. Very possibly, all you could be justified in claiming is that among the many elements of which you are to various degrees conscious at any given moment there is at least some central group that is known as an existentially unified whole. There could be plenty of room for disagreement here. Yet this much, at any rate, has been very widely accepted: that at least some ingredients of our universe, namely, conscious lives or particular conscious states, or particular groups of elements inside such states, can be known immediately (that is, just through being experienced) to be wholes whose constituents are abstractions, not separately existing entities. The suggestion that human consciousness is itself always just a matter of *separately existing whatnots arranged in various spatial patterns and causal sequences* has been rejected by one after another of a long line of thinkers.

The theme is found in Plato's writings. In the *Phaedo* we learn that the existence of a human soul involves more than just 'a harmony', a structural unity. If souls were merely elegantly functioning collections of separately existing elements, then they could fall to bits in due course. Since, however, they are not, each soul's immortality is assured. Again, Descartes wrote in his *Sixth Meditation* that 'the mind is very different from the body, because when I contemplate my mind—myself, in other words, in so far as I am merely a conscious being—I can distinguish no parts'. He noted that his mind was a complex arena of sensations, emotions, acts of will and of understanding, and so forth, but to him these were self-evidently not 'parts' in the philosophically traditional sense of being entities each genuinely separate in its existence.

Locke expressed similar convictions in his *Essay Concerning Human Understanding*. To picture God, an eternal thinking being, as 'nothing else but a composition of Particles of Matter, each whereof is incogitative', would, he said, 'ascribe all the Wisdom and Knowledge of that eternal Being, only to the juxtaposition of parts; than which, nothing can be more absurd'. He viewed his point as applicable not merely to the divine mind but to all other minds as well. 'For unthinking Particles of Matter,' he commented, 'however put together, can have nothing thereby added to them, but a new relation of Position, which 'tis impossible should give thought and knowledge to them' (1700: book 4, ch. 10, s. 16). Leibniz dreamt up a vivid means of conveying the point. Consciousness, he wrote, could never be explained '*on mechanical principles*, i.e. by shapes and movements. If we pretend that there is a machine whose structure makes it think, sense and have perceptions, then we can conceive it enlarged so that we might go inside it as into a mill', but then 'we shall find there nothing but parts which push one another, and never anything which could explain a perception. Thus, perception must be sought in simple substance, not in what is composite (1714: *Monadology*, s. 17).

This way of thinking was so common that Hume could write as follows in his *Treatise on Human Nature*:

I observe first the universe of objects or of body: The sun, moon and stars; the earth, seas, plants, animals. Here *Spinoza* appears, and tells me, that these are only modifications; and that the subject in which they inhere is simple, uncompounded and indivisible. After this I consider the universe of thought, or my impressions and ideas. There I observe another sun, moon and stars; an earth, and seas, plants and animals. Theologians present themselves, and tell me, that these also are modifications, and modifications of one simple, uncompounded and indivisible substance [his immaterial soul, that is to say].

'Immediately upon which,' he exclaimed gleefully, 'I am deafen'd with the noise of a hundred voices, that treat the first hypothesis with detestation and scorn, and the second with applause and veneration (1739: book 1, part 4, s. 5). Gleefully because he opposed both hypotheses and was trying to profit from the hatred his contemporaries felt for what he called 'all those sentiments, for which *Spinoza* is so universally infamous'. The kind of unity that Spinoza attributed to the entire universe was thought by almost everybody to characterize human minds. It was a unity cementing together all the models of external objects (sun, moon, stars, etc.) that were present in any mind when it thought about such objects.

Bradley could be the philosopher who gave most thought to the topic. 'Every relation, to be possible, must itself bear the character of an element within a felt unity', he declared; things are linked in 'a mode of togetherness such as we can verify in feeling', one which 'destroys the independence of the reals', showing them to be joined in their being; 'I fail to understand the position of those who seek apparently to deny or ignore the very existence of what I call "feeling"—an experience, that is, which holds a many in one, and contains a diversity within a unity which is not itself relational'; 'immediate experience gives us a unity and unities of one and many', and to attempt to treat all such unification 'as consisting in no more than some relation or relations, I cannot but regard as really monstrous' (Bradley 1893: 125; 1914: 281; 1935: 633, 634, 663). But it is doubtful that all this could give us much insight into the affair. How could elements of a conscious state manage to be more than separately existing things glued together by such relationships as A causing B, or C being near to D in space or in time? Bradley confessed that this was 'not completely analysable'. William James may thus have achieved as much when he wrote simply that, 'however complex the object may be, the thought of it is one undivided state of consciousness', or C. D. Broad when he remarked, 'I doubt whether anyone except a philosopher engaged in philosophising believes for a moment that the relation of "himself" to "his toothache" is the same relation as that of the British Army to Private John Smith' (James 1890: 276 of Dover repr.; Broad 1925: 585).

Such views are echoed by numerous writers of recent times. J. L. Mackie points to 'experiential content' as being 'the one feature of our mental life that we would hesitate to ascribe to a computer, however sophisticated its performance', for 'the basic fact of occurrent awareness seems not to be analysable into any simpler components' (1982: 127). Abner Shimony considers that when a theory describes the universe as a collection of entities never entangled with one another with respect to their actual being, then that theory will be hard to reconcile with 'the holistic character of mind that our high-level experience reveals'.[7] Ian Marshall speaks of the problem presented by 'the unity and complexity of states of consciousness', their 'diversity-in-unity' which must surely be grounded in something 'unanalyzable into parts with separate identities'(1989: 73 and 79); David Hodgson of the need to explain how 'large amounts of information seem to be available *all at once* in conscious experiences, without any necessity to scan' (1991: 110); and Roger Penrose of the puzzle presented by 'the "unitary sense of self" that seems to be a characteristic of consciousness', the 'oneness' or 'globality' of consciousness, the strong contrast between it and any collection of 'a great many independent activities going on all at once' (1994: 368; 1987: 274; 1989: 398–9). Michael Lockwood's *Mind, Brain and the Quantum* is almost entirely concerned with this puzzle, while the title chosen by Penrose for his most famous treatment of the area, *The Emperor's New Mind,* expresses his dismay at typical modern efforts to overlook it.

How it Feels to Be a Conscious Being

As Lockwood insists, the idea that a conscious state is something 'that is experienced as a whole' does not deny that consciousness is

---

[7] Shimony 1997: 149; pp. 150–6 suggest that quantum entanglement could help solve the problem.

complicated so that a unified experience 'can have experiences as parts'. 'If', he comments, 'I see a woman standing by a horse, I have a visual experience which contains as parts the experience of seeing the woman and the experience of seeing the horse' (1989: 87). A conscious state can be made up of a vast number of elements that are in some important fashion separate from one another. They can stand to one another in complex relationships: relationships of being close together or far apart, of surrounding or of being in between, of forming round shapes and others that are square, and so on. Here, the terms 'close', 'far', 'surrounding', etc., no doubt carry rather a different meaning from when they are applied to things in physical space. It is instead the space (or, if you insist, *quasi-space*[8]) of experience, phenomenal space, that is here in question. But the fact that such terms are applicable does show that conscious states can *have parts* in some perfectly respectable sense, although Descartes and others prefer not to use any such language. The crucial question is whether the distinguishable elements of such states are unified in their existence. If we agree that they are, then whether to call them 'parts' becomes a trivial matter of verbal preference, I suggest.

It might be best to take a similarly tolerant line when asking

---

[8] Do ordinary ways of talking demand that the word 'space' be applied to physical space only? Surely not. People find it natural to use such words as 'round' and 'square' to describe afterimages of bright lights, for example. Now, how could anything be describable as 'round' without being spread out in something which had at least some slight claim to be called 'space'? Nothing odd is proposed here. It is not as if one were asked to believe that the real world, in addition to having elements spread out in physical space, included *a second set of elements* spread out in another space. Instead, there could be elements which were distributed in physical space, inside the brain, *and which were also* arranged in phenomenal space, the sheer fact that two elements were close together in phenomenal space not dictating that they had to be close together in physical space as well. Cerebral activity-patterns representing a chess king in check from a pawn—so that it was right next to the pawn on the chessboard—might occur in the brain's left hemisphere while those which represented the pawn itself were occurring in the right hemisphere. (If two of its activity-patterns seem to a thinking, perceiving brain to be close together, then they are *ipso facto* close together in phenomenal space, alias perceptual or mental space, a space in which they appear without, of course, the brain therefore being aware that cerebral activity-patterns are what they are.)

whether advanced computers should be called conscious of their surroundings or of the intricate events inside themselves. Imagine a computer able to compose what looked like good poetry, to drive a car down a busy road, and to give an account of the steps it went through while working out how to perform these feats. Denying that it was in any useful sense conscious might be considered almost as odd as saying it was in no sense good at chess although able to thrash grandmasters. At any rate I myself would see little point in it. The crucial matter is instead that such a computer might have no conscious states of the kind believed in by Plato, Descartes, Locke, Leibniz, Bradley, James, Broad, Mackie, Shimony, Marshall, Hodgson, Penrose, and Lockwood, states containing elements that were existentially unified. Now, while states that lacked elements unified in this way could none the less have some right to be labelled 'conscious states', they would (if my earlier arguments were correct) lack all intrinsic value, and they wouldn't be conscious states of the kind I believe are had by you and me. They therefore wouldn't be cases of what I'd happily call 'full consciousness'—and if you wanted to say they weren't cases of 'consciousness in the fullest sense', or even of 'consciousness in the ordinary sense', then I wouldn't protest much.

Was John Searle right to be indignant at the idea that 'if you made a computer out of old beer cans powered by windmills, if it had the right program, it would have to have a mind'?[9] Was Ned Block (1978) correct in arguing that, if all the people of some vast land of the future—some super-China—could be organized so that they functioned collectively like the tens of billions of cells in a human brain, then the result still wouldn't be a collective mind? I suggest that whether to use the word 'mind' in such cases isn't the really interesting issue. The key question is instead this. Would the beer-cans computer and the super-Chinese collectivity have states sufficiently unified to have intrinsic value, states of the kind we detect when we experience our own mental workings?

[9] Searle 1984: 14 of the Second Lecture ('Beer cans and meat machines', 15 November, pp. 14–16).

Granted that I view our entire universe as a Spinozistic whole, its parts united in their existence, wouldn't my answer have to be 'Yes, they would'? Not necessarily. The universal whole in which I believe contains vastly many parts. The sheer fact that various of these parts *were all of them ingredients of this whole* would presumably not be enough to guarantee that they formed a subsystem possessing intrinsic value. The seven pencils on my desk, for example, presumably wouldn't form such a subsystem. Further, it isn't clear that merely linking together a huge collection of pencils so that they performed complex computations, perhaps even composing poetry, could make them into such a subsystem. Shimony, Marshall, Hodgson, Penrose, and Lockwood all suggest that elements must be unified *in particularly thoroughgoing ways which are familiar to quantum theorists* if they are to enter into conscious states of the kinds had by you and me. Now, they may well be right.

Imagine a computer working on much the same lines as the one on my desk but a trillion times more powerful. The computer might well discuss hard philosophical issues far more clearly than I can, but would there be, in the well-known phrase used by Timothy Sprigge and later by Thomas Nagel,[10] anything *which it was like to be* that computer? I very much doubt it. What's in question is whether the computer could know its own states as you and I know our conscious states. Suppose that the computer, besides discussing philosophy expertly, actually managed to report on its internal workings in immense detail. This still wouldn't settle the question, as numerous philosophers have pointed out. People are right to view introspection with some suspicion but its findings mustn't be dismissed wholesale, and the following are among the most important

---

[10] Sprigge 1971: 167–8; Nagel 1979: ch. 12 ('What is it Like to be a Bat?', originally published in *Philosophical Review,* Oct. 1974). See also Searle 1997: 201: 'Is there something it is like, or feels like, just to sit and consciously think that 2+3=5? And if so how does that differ from what it feels like to sit and think that the Democrats will win the next election? There is indeed something that it is like, or feels like, to think these things, and the difference is precisely the difference between consciously thinking "2+3=5" and consciously thinking "The Democrats will win the next election".'

of them. First, our experiences have *qualitative aspects* (experienced blueness, sweetness, etc.) of kinds that seemingly couldn't be reproduced by any mere system of electronic impulses (or beer-can movements, or whatever) developing inside anything that was in the least like my computer in its operations. Second, these experiences also have *unification* of a kind going beyond any mere structural unity—any complex variant on the unity of an army of ants working in orderly ways. Such findings may be difficult to handle philosophically yet that is no excuse for rejecting them. The next sections will look closely at this area.

Knowledge beyond Mere Knowledge of Structure

Suppose various ripe oranges are all experienced as of the same colour. It strikes me as peculiar to claim—as would Daniel Dennett (1991: ch. 12)—that the colour-experience in itself, as distinct from any associations and emotional reactions coming with it, amounts to nothing more than could be had by just any computer of today which, hooked up to a TV camera, was able to sort fruit according to their colours. Admittedly, talk of 'the indescribable what-it's-like-to-be-conscious-of-colours-and-shapes' will not help blind people towards an understanding of precisely how it feels to experience various cherries and apples as all of them red and round, various bananas as all of them yellow and elongated. Still, sighted people can at least know they possess far more than a mere awareness that the fruit of the one set fall into one classification, the fruit of the other set into another, with those in the first classification arousing such and such a group of dispositions to react while those in the second arouse such and such another group—because this is the sort of awareness enjoyed by 'blindsighted' people, victims of brain injury who insist that their visual ability involves experiences altogether different from those of ordinary vision. Possessors of blindsight can often distinguish reliably, on the basis of their shapes, between bananas and apples when these are presented in areas to which they say they are blind; they may actually be able to classify objects in such areas according to their colours;[11] but they have the impression of being engaged in mere guesswork. When their blindsight is at work, they report, they entirely lack the experiences of shape and of colour that humans normally have and that they themselves continue to have in the cases of any areas of normal vision that exist side by side with the areas they scan by blindsight only. Mere introspection is enough to tell them this.[12]

Similarly with the kind of unification we are aware of at particular conscious moments. Mere introspection can establish that the unity of elements forming a simple experience—the experience, say, of three points of light closely grouped in what is otherwise total darkness—isn't a unity of things each having an existence truly separate from that of the others. On this sort of point Descartes was right. Introspection can reveal a unity clearly different from any unity to be found in the quasi-thinking and quasi-experiences of what President Eisenhower termed 'the military-industrial complex'. The military-industrial complex is a whole operating much like a human mind, you sometimes hear. It has a quasi-cunning permitting it to produce more and more armaments, seemingly regardless of whether the humans who compose it—the individual cells, so to speak, of its gigantic brain—are in

[11] See Stoerig and Cowey 1989. Dennett knows all about blindsight, actually citing Stoerig and Cowey (Dennett 1991: 325). Lockwood comments that Dennett's position is in effect that normal sight 'carries with it far greater confidence in the corresponding judgments, and is of vastly greater discriminative power', but that there is no radical distinction between it and blindsight, people's reports to the contrary showing only that they are victims of 'sheer illusion' (Lockwood 1993: 68; cf. Dennett 1991: 374, which goes so far as to tell us that between a simple machine's colour-discrimination and a human's there is 'no *qualitative* difference').

[12] Mackie 1976: 167 rejects J. J. C. Smart's theory that knowing what it's like to experience a particular colour is simply knowing that *something is going on inside oneself which is like* what is going on inside oneself at other times when the same colour is experienced. As Mackie notes, the difficulty is that it certainly seems to us that we are aware of something more specific. Instead of just knowing that similar things are going on inside us when we experience *orange afterimages* on the one hand, *actual orange patches on walls* on the other, 'we know the respect in which' the experiences are similar; we know 'a certain phenomenal quality, the colour as seen'. See Dennett 1991: 389 for a defiant declaration that while it does indeed seem that knowing how it feels to experience a colour is more than simply being aware of some package of personal dispositions to react—for instance by saying 'What a beautiful shade of orange!' in response both to an afterimage and to a patch on a wall—'this is just how it seems to you, not how it is'.

favour of arms control. But its substitute for an ordinary mental life is completely lacking in what makes ordinary mental lives worth living. The military-industrial complex is far too much *split up into separate entities* to possess what many of us would be eager to call 'conscious states'.

The same would be true of any system consisting of hugely many men who passed to one another slips of paper each bearing a zero or a one, in compliance with a set of rules—'a computer program' we could say—of which no man knew more than a tiny fragment. (One of the men could perhaps know nothing beyond the rule 'Pass a zero to the next man when given a zero and a zero'.) The system as a whole might beat grandmasters at chess, compose great poetry, discuss philosophy of mind brilliantly, answer questions in Chinese whereas all the men in the system spoke nothing but English,[13] and—when hooked up to TV cameras—describe colourful paintings as skilfully as any art expert; nevertheless, any understanding the system possessed of chess, poetry, philosophy, Chinese, or the beauties of colour wouldn't be understanding *in itself worth having,* I'd insist.

I'd insist on it despite my Spinozistic belief that not just all the men in any such system, but absolutely all the things of our universe, are mere aspects of an existentially unified whole. Inside any greater unity possessed by our universe in its entirety there are complex systems whose elements are still more thoroughly unified: unified in ways fundamental to any intrinsically worthwhile experiences of ours. Knowledge of our own conscious states shows us that they are instances of systems whose elements are unified in these ways, whereas knowledge of collections of beer cans, transistors, men passing slips of paper to one another, and similar 'hardware' on which what might be called 'computer programs' could be running, indicates that they lack such unification.

Some people cannot detect redness (protanopia) while others cannot detect blueness (tritanopia). Imagine a philosopher blind from

---

[13] Searle is famous for his discussions of a 'Chinese Room' in which a man ignorant of Chinese goes through a long series of rule-controlled steps which, unbeknownst to him, produce answers in Chinese to questions posed in that language. See e.g. Searle 1997: 11.

birth onwards who becomes immensely impressed by this fact. Our philosopher develops the hypothesis that human minds are never immediately aware of any such reality as experienced purple, *phenomenal red-blue*. Instead, a human viewing a purple object has an experience of something as possessing the quality of reddishness, *and has in addition* other experiences of the same thing as possessing the quality of bluishness, as being cold, as being heavy, as making a sound like a lovesick turtle, and so forth. While the experiences are very closely linked, they nevertheless exist separately from one another, says the hypothesis. How ought we to refute it? Not by pointing out that the words 'experienced purple' are ordinary English: the matter in dispute isn't how English functions, but how our minds do. Now, what introspection in fact reveals is that experiencing purple definitely is not a case of having two separately existing experiences. There is no need to ask neurophysiology to prove this. It is how states of mind *are to themselves,* the what-it-feels-like-to-be-in-these-states, that is in question, rather than how any set of active brain cells or anything else looks when scientists examine it from outside. Introspection is all we need in order to find whether our blind philosopher is mistaken. True enough, introspective reports are often faulty because, for example, of all the motives people have for self-deception and all the problems of describing situations clearly—yet on the question of whether reddishness and bluishness are always experienced separately, instead of *red-blueness* being experienced as a completely unified whole, introspection is straightforward and decisive. And similar remarks apply, I suggest, to what introspection reveals about the unification of various elements inside fairly complicated conscious states, although here the unified elements will *very much more clearly deserve to be called 'parts'* than the red 'part' and the blue 'part' of experienced red-blue. Think once again of the consciousness of the three points of light in what is otherwise total darkness. Is this an experience of the first point of light, plus a second experience which is of the second point, plus a third experience which is of the third, these three experiences being no more than *closely associated* so that it is only courtesy of ordinary English, notorious for papering over many a gap, that we can talk of 'a single experience of three points of light'? Would people from Plato to Lockwood all be talking nonsense when they theorized that something beyond structural unity was involved in any case of this kind? It strikes me as plain enough—just from introspection—that the answer has to be 'No'.

Partial Fusions could Explain the Qualitative Nature of Consciousness

It is because various mental states *are more than mere complex groupings of active brain cells* that the region of one's mind that is conscious of various things and then formulates reports on them can be a region unified in a way enabling it to report that an experience of two fruit, for instance, as being identical in colour, isn't something such as could be had by just any machine able to sort fruit according to their colours. The region's unity allows the region *to know itself* not simply in its structure but also with respect to the qualities which make the difference between something as abstract as structure—organization of the kind had by sets of impulses originating in colour-sensitive electronic devices and processed in accordance with various computational rules—and any concrete reality.

As Lockwood notes, while physicists describe the material world in terms of its spatiotemporal structure they seem never to get around to telling us 'how the structure is qualitatively fleshed out', yet surely 'it must be fleshed out in some way or other'. If a physicist's description of the world's structure is to be true, then the structure 'must be concretely realized; the physical world must have an intrinsic nature which instantiates the structure in question'. The physical world cannot be just a matter of structure, structure, all the way down! Accepting what he takes to be a theory rather poorly expressed in Russell's *The Analysis of Matter*, Lockwood argues that in consciousness the intrinsic, qualitative nature of the physical world, or at least of some of it, 'makes itself manifest'. How? He answers that our conscious states are physical states of kinds that have *a unity that allows them to know their own qualitative natures.* The clue to this, he thinks, is that quantum theory permits wholes to have parts whose identities *are partially fused.* (Remember the two bosons in the box.) This is why the wholes can know not just their own intricately organized structures, but also the nature of the stuff from which those structures are built.[14]

Whether or not quantum theory really could solve the puzzles of this area, the right approach to them must presumably be on much the lines that Lockwood indicates. The difficulty is in seeing how the stuff of any intricately structured mental state could manage to be stuff grasped in itself, understood with respect to its own qualities, rather than merely being identified as the something-one-knows-not-what which forms various structures that have come to be known. Knowledge of structure is fairly easily had, so long as we are fairly liberal about what can be counted as knowledge. Any typical modern computer can readily be programmed to print reports on (and therefore to 'show knowledge of', in some suitably weak sense) various structures which have been built up inside it, intricate patterns arrived at through skilful integration of hugely many items of information. But now consider consciousness of afterimages produced by bright lights. Just what is it like to experience the redness of one afterimage, the blueness of another? This is not mere knowledge of structure: knowledge such as could in theory be conveyed by a lengthy telephone call to someone blind from birth onwards who could then exclaim, 'I now understand exactly what it must be like to experience the first afterimage as red and the second as blue.' It cannot be a matter just of knowing information which has been processed in a brain until it forms a pattern of much the kind that a modern computer builds up by shuttling impulses from place to place. One's consciousness of the afterimages—or, if you prefer, one's mind when it is conscious of the afterimages, or some region of one's mind that is conscious of them—forms a whole that includes knowledge of more

---

[14] Lockwood 1989: 238. See also, in particular, p. 159 which considers Russell's apparently preposterous suggestion (1927: 320 and 383) that a physiologist always really sees events in his own brain, never ones in the brain of another person. Lockwood views this as an unfortunately phrased expression of the idea that there are 'radically different ways in which the very same brain events might be known—on the one hand by perception, aided perhaps by modern instruments; and on the other hand by self-awareness: knowing certain brain events by virtue of their belonging to one's own conscious biography, knowing them, moreover, in part, *as they are in themselves*—knowing them "from the inside", by living them'.

than just *the arrangement of its parts*. Now, Lockwood would seem right in his hunch that it could do so only by knowing itself as a whole inside which the identities of the parts were in partial fusion.

Why say this? If the identities of its parts were partially fused, then how could this help anything to know itself as more than just a complexly structured set of elements whose own qualitative natures were unknown? We can answer for a start that the qualitative natures in question, were they to be known at all, could presumably not be known through the individual atoms or electrons of a brain, or maybe its individual cells, *themselves knowing their own qualities, then passing their knowledge onwards* so that it could be grasped by consciousness. Rather, the knowledge would have to originate at a higher level through an entire state of mind—or at least a sizeable part of it—knowing itself as a whole. Still, when the whole knew itself as a whole, just how might the fact that the parts 'had partially fused identities' contribute towards any acquaintance with the intrinsic qualities of the stuff of which the whole was composed? How could a whole, when its parts were as closely united as (for instance) bosons in the same quantum state can be united, come to know itself in a fashion radically different from that in which a computer of today could come to know itself, so that instead of possessing simply the kind of knowledge that such computers build up through the interactions of their transistors, knowledge of structure, it came to have direct knowledge of the stuff from which its structure was made?

Unable to offer a detailed answer, I can at least make the following point. Having parts with partially fused identities would mean that a complex conscious state, one that had knowledge of the many parts of which it was composed, would have knowledge *not necessarily restricted rigidly* to knowledge built up by the causal interactions of many separate entities, as in computers of today, since the causal interactions between its parts wouldn't be all that brought those parts into the whole. (The whole wouldn't simply be various things plus their causal interactions, the things counting as parts of this whole purely because of how they interacted. Fusion of identities as seen in bosons occupying the same quantum state is no mere matter of close interaction, like the fusion of individual buffaloes into a stampeding herd.) While you could still perhaps believe in the rigid restriction in question, it at least wouldn't be an utterly inevitable restriction. It wouldn't be present with a necessity that was obvious as soon as you considered the affair.

The puzzle posed by our knowledge of phenomenal characteristics, qualities such as the redness or blueness of afterimages, is known as 'the problem of *qualia*'. Qualia include such affairs as the sound of a heard note, the painfulness of a sensation, the taste of a wine; possibly, too, how it feels to be in love or just what it's like to understand that two and two make four. Well, philosophers need to take qualia very seriously, I'd say. But this isn't the same as declaring that my mental life would cease to be worth living if all my knowledge of qualia were replaced by mere knowledge of structure: knowledge like an immensely rich variety of blindsight. To judge by their writings, the mental lives of some philosophers include nothing more than that, and they might be very fine mental lives all the same.

Also, I don't think we are guaranteed against ever being mistaken about our qualia. I recognize that we are often very bad at reporting our own conscious states. (Suppose an object is presented near the edge of your visual field as you stare firmly ahead of you. You may find yourself quite unable to say whether it is red or blue, and this may well come as a big surprise to you.) Again, when I know the difference between experienced redness and experienced blueness, what I know is a difference as it appears in my consciousness at the present moment. For all I know and for all I care, some wonderful contraption may be beaming waves at my head so that every so often all my colour qualia are altered, simultaneously tampering with my memories so that the alterations never become known to me. My interest in qualia is almost entirely due to the fact that they seem to be sure signs that in conscious states at particular moments there can be unification of a kind never found in any ordinary computer. Qualia show that my mind's knowledge of itself goes beyond anything that could be had by any system of vastly many people passing slips of paper to one another, or of transistors organized much as those sitting on my desk are organized.

Not one of these arguments depends on tyrannical standards of what can be counted as *knowledge* or as *self-knowledge*. I fully admit that there are kinds of colour-knowledge that might be had by a system of old beer cans suitably linked by rods and cogwheels, supplied with an appropriate computer program, and then fed with impulses from colour-television cameras. The beer-cans computer might readily identify fruit as red or yellow. It might further describe its internal model of the fruit, plus its method of remembering that all the red fruit fell into one category, all the yellow fruit into another. All this, no doubt, would be knowledge of colour of some sort, and self-knowledge of some sort: knowledge of the colour of the fruit, and also what we might conceivably want to call 'knowledge of the phenomenal colouring of the computer's internal models of the fruit'. But in what form would the computer possess its *knowledge of phenomenal colouring* (if, bending over backwards in our eagerness to avoid linguistic tyranny, we agreed to call it that)? It would possess it as knowledge of the structure of various of its own states. It would be knowledge such as could be acquired by any blind-from-birth computer expert who was given enough information about how at various times the beer cans were positioned relative to one another. This wouldn't be knowledge of phenomenal colouring of anything like the sort that sighted people can enjoy when they know their conscious states. Being held together by rods and cogwheels cannot unify beer cans into a whole of the right variety.

There is a sense in which a beer-cans computer would be a whole 'only in the eye of its beholders', rather as a forest is, the underlying reality being the individual trees or perhaps the individual quarks and leptons that make up the atoms that make up the trees. But, once again wishing to avoid linguistic tyranny, let me insist that the beer-cans computer would *in another, weaker sense* be a whole *in itself*—even a pile of sand forms a whole in a weak sense, doesn't it?—and that it might actually be a whole that in some sense got to know the stuff from which it was made. It could, that is to say, receive and understand, in the perhaps very weak sense in which present-day computers can understand things, a metallurgist's report on the alloys in its cogwheels and beer cans, and a physicist's report on the neutrons, protons, and electrons inside the alloys, and a further report on the quarks inside the neutrons and protons. But the crucial point is that these reports would all be in terms of structure, structure, and yet more structure—no qualia there!

To Varying Degrees, Conscious States can be Split up, and Quantum Theory might Say Why

Ability to know qualia depends, I have been arguing, on how one's consciousness at any given moment can have elements that are unified in a striking fashion, elements whose identities are in partial fusion. Quantum theory might well be relevant, as might the Spinozistic idea that a highly complex whole can be unified in its existence, its parts being abstractions much as the length of a lake or its blueness are abstractions (while of course remaining fully real). Thinking that various elements have 'at least partially fused identities' involves viewing each as an abstraction from a whole formed by it and the other elements. What is more, there is no conflict between saying that parts of some conscious state at a particular moment *form an existentially unified whole* and declaring with Spinoza that our entire universe is such a whole; for why on earth should holding that items *A, B,* and *C* are all unified in their existence involve a denial that *D* and *E* and *F* and *G* are unified with those same items and with one another in the same manner? Existentially unified wholes could sometimes be mere constituents of greater existentially unified wholes.

They might, too, be wholes whose unification was more impressive than that of the greater wholes into which they entered. Their elements might be to various degrees more closely integrated. True enough, the greater wholes would be ones in which every element was linked in its very being to every other element, but inside the smaller wholes the linkage could be more obvious. Again, the smaller wholes could include yet smaller ones, and inside those it could be still more obvious. This might be of great significance for our conscious lives.

Let us look at such points in more detail. For a start, some complex quantum phenomena are 'holographic': the characteristics of the individual elements in the complexity are to some extent distributed throughout the whole. As several physicists have noted, this might help us understand the nature of consciousness. Bohm and Basil Hiley write that in a hologram of an object 'each region makes possible an image of the whole object'; 'in some sense, the whole object is enfolded in each part of the hologram'. Now, they remark, similar things could be said of consciousness 'with its constant flow of evanescent thoughts, feelings, desires, urges and impulses' which all of them 'flow into and out of each other and, in a certain sense, enfold each other'; 'our most primary experience in consciousness' is of this kind of enfoldment, they state. (One of Bohm's favourite stories is of how an ink blob inserted in glycerine between two cylinders becomes thoroughly spread out when one of the cylinders is rotated yet can be recovered intact—'unfolded' or 'made explicate' when it was previously 'enfolded' or 'implicate' throughout a large volume of the glycerine—by reversal of the rotation.) Bohm and Hiley tell us that quantum theorists are continually finding enfoldment, 'implicate order', of one sort or another. 'A rudimentary mind-like quality is present even at the level of particle physics', they declare; there is 'a basic similarity between the quantum behaviour of a system of electrons for example and the behaviour of mind' (1993: 353–4, 382, 385–6).

Danah Zohar, too, finds holograms highly suggestive, commenting that they 'are in fact quantum structures'. They are 'pictures "written on" laser beams' and the photons in these beams have partially lost their separate identities through being forced into the same quantum state. Holograms may therefore offer us 'important models for the mind because they address the unity question'. In a hologram, the whole 'is represented in every part' and this 'seems to reflect the reality of our conscious life' (Zohar 1996: 446).

In quantum phenomena, though, being distributed throughout the whole turns out to be very much a matter of degree. Remember, physics tells us that because of quantum entanglement every particle is linked with all the other particles with which it has ever interacted, and hence also with all the further particles with which those in turn have interacted. From this it follows that, of all the particles in all the galaxies visible to our telescopes, at least a large proportion (and perhaps the lot) are joined up into a whole whose every part has some continuing connection with every other. Still, isn't it plain that for all practical purposes many of the parts are only very minimally connected? Perhaps only a computer the size of a planet and fed with immensely many details of events from the early Big Bang onwards could form anything like a detailed understanding of how this works out in practice, but the situation's general outlines can be described. (i) The constituents of our world, considered each in isolation, are often or always abstractions in something like the way in which an object's size or its mass is an abstraction, yet (ii) this is far from evident at first glance, and (iii) the reasons why it isn't evident can be explained, but only very roughly, by the quantum theorists of today. (Why do the entities that common sense classifies as 'individual things' appear so distinct from one another? How, to put the point technically, did the universe's quantum wave function come to suffer decoherence as the Big Bang cooled? Physicists such as Murray Gell-Mann and Jim Hartle have at least begun to reply to these questions. Part of the answer may well be that there are valid ways of conceiving wave functions which make their collapses into definite sets of particular events *somehow relative or even to some extent fictitious*, as is suggested by what has come to be known, perhaps confusingly, as 'many-worlds quantum theory'.)

Similar comments apply to what introspection reveals. While introspection indicates that conscious experiences can be wholes of a dramatically unified kind, wholes whose wholeness is grasped by the wholes themselves, it is plain that not everything of which anybody is conscious at any point in time is grasped with the same degree of firmness. You can be only half conscious of various things, while sometimes 'half conscious' can look altogether too generous a description. Psychologists can trick you with a computer screen which seems to show an unchanging set of words when in fact it is seething with activity. The trickery involves changes to the screen whenever your eyes move, thanks to movement-detectors that trigger rapid computational processes. In the moments when your eyes are jumping towards new words, these are replaced by others of the same lengths. While other people see the screen as a wriggling mess, you think you are viewing a large and unvarying area. Although any impression that every bit of it is detected at high resolution is an illusion, the large area is all of it detected visually. The changes to it are simply *not noticed,* just as the fact that the edges of your visual field are experienced very poorly (so that the colours of any objects appearing there can often only be guessed wildly) is typically not noticed—because when you want to take note of an object your eyes usually turn automatically so that you can view it at high resolution by looking straight at it.

Pantheists need find none of this troubling. Was Bradley correct when he wrote (1914: 346) that Reality was ultimately 'a single Experience'? Even if he was, the cosmos wouldn't have to be self-knowing in so thorough a manner that its every part was vividly aware of every other part so that it became what James called 'a large seaside boarding-house with no private bed-room in which I might take refuge from the society of the place' (1912: 277). Something unified in its existence, its parts mere abstractions rather as a lake's length is a mere abstraction, could still have greatly varied parts. In the case of existentially unified wholes possessing consciousness, one respect in which the parts might vary would be the extent to which each was ignorant of the natures of the others, the ignorance perhaps sometimes actually extending to the fact that the others existed. Inside any such whole— even Bradley's 'single Experience'—there could be many smaller wholes which grasped one another's characters with greatly varying degrees of success. Some of them could be human conscious lives, and even inside these there could be further wholes, once again varying in the awareness that each had of the others.

Again, it could be that elements *A–B–C–D* formed an existentially unified whole, and so did elements *C–D–E–F*. The two wholes could then be said to overlap in the region *C–D*. The idea of overlapping has been investigated by Lockwood. He rejects the assumption (shared, he says, by Descartes and the man in the street) that every conscious state has a unity which is absolute in two ways: first, that all its parts enter into it with the same clarity, and second that the division between this particular conscious state and any other is a hard and fast division. In point of fact, he writes, there can be 'overlapping phenomenal perspectives' with 'a spectrum of degrees of connectedness'. The difference between a normal mental life and the startlingly disunified mental life of someone who has undergone an incomplete commisurotomy—somebody, that is, whose cerebral hemispheres have become largely disconnected through cuts made in the thick bundle of fibres joining them—can be considerably less than we tend to believe. Various regions of conscious lives are often joined only extremely indistinctly. Region #1 may overlap very partially with region #2, which may in turn be connected in the same weak manner to region #3, the first region and the third then being linked only very weakly indeed. While in healthy people at particular conscious moments there may be 'a large number of phenomenal perspectives which approximately coincide', brain damage can result in a significant 'spread' or 'dissociation' between such perspectives. But a fuzzy kind of unity, a unity based on overlaps, can still manage to be a unity experienced to some degree.

All this might help explain why states of awareness seem to cover brief temporal intervals, as when musical notes played in swift succession appear to be experienced together yet with the earliest ones fading away from consciousness. To many philosophers such a phenomenon could seem 'a *reductio ad absurdum* of the notion of degrees of co-consciousness', Lockwood notes. These philosophers believe that all that ever really exists is the situation of a present moment that has absolutely no temporal spread. According to their theory the future is completely unborn, *not real yet*, and the past has vanished utterly. Events of earlier and later times are *non-existent* in a way that isn't purely relative, they think. The non-existence is no mere matter of being elsewhere along a time dimension (relative to us now) much as people in Argentina are (relative to people in China) elsewhere in space. On such a theory, Lockwood comments, any supposed unity between past and present, any so-called 'temporal extendedness within a phenomenal perspective', must be 'an illusion generated by the simultaneous co-presence within consciousness of current experiences and what, strictly speaking, are merely vivid *memories* of past experiences'. However, he follows Albert Einstein in rejecting the theory in question.[15]

---

[15] My Chapter 1 quoted Einstein as concluding that his theory of relativity made it 'natural to think of a four-dimensional existence instead of, as hitherto, the evolution of a three-dimensional existence'. Lockwood's words are from 1989: 89–94 and 98–9.

Overlaps could be Crucial to Perceptions and Thoughts that are Intrinsically Worth Having

It is tempting to speculate that our brains, building up their internal models of the world, make use of overlaps in representing such matters as *being of the same colour*. In fact, we might find here some limited vindication of a curious theory perhaps originating in Plato's difficult remarks about the Forms, and suggested by things said by Hegel and his successors: things leading to their being accused of 'confusing the *is* of attribution (as in "Sugar is sweet") with the *is* of identity (as in "Bonaparte is Napoleon")'. The theory is that any similarity uniting various entities is always just a matter of those entities being fused in a particular way. (In his marvellous story *Tlön Uqbar Orbis Tertius,* Borges records that one of the sects of Tlön defends 'the Platonic idea that all who repeat a line of Shakespeare *are* Shakespeare'.) Such a theory could be on the right lines to at least this extent: that when, say, two afterimages seen side by side were experienced as both of them purple, then overlapping of the kind I have been discussing would be crucial to the appreciation of their common purpleness.

This might make such overlapping central to various thoughts of a primitive type. The notion that there could be consciousness of the two cases of purpleness without the slightest grasp of their similarity strikes me as a mistake; and what is a grasp of a similarity if not a simple thought, 'There's similarity here'? Animals lacking any form of language could still grasp similarities, but I do not view language as an ingredient in every thought—which is one reason why I believe that human babies, and horses, fish, and frogs, often have lives well worth experiencing. You could continue to think 'Those things are similar' after brain damage had destroyed your knowledge of *the word for* similarity.

Again, consider what is going on when you appreciate that a greyness belongs to a large stone. How is this managed? As Peter Forrest writes, 'Perceptual sensations *represent,* and it is not just that they can represent given a suitable interpretation—anything can represent anything given a suitable interpretation. Rather, they carry their interpretation with them: *they are intrinsically meaningful.*' Still, how do they achieve this? I suggest that your visual image of the stone is combined with many further images derived from past experience, images of hardness, heaviness, etcetera (for the word 'images' can be applied to imagery which is not visual). The result could be considered a thought of a more complicated kind, expressible by words like 'Here is an object not only grey but also hard and heavy'. On this kind of point, various philosophers of the seventeenth and eighteenth centuries seem to me right, although their views about the nature of images may often have been rather too simple. Those of today who teach instead that the term 'thoughts' must always stand for *dispositions to behave in particular ways, combined with strings of words* (imagined if not actually spoken), strike me as picturing our mental lives as very little worth living. And here, once again, the idea of overlapping might be useful. It might help us to grasp what went on inside brains when things were appreciated as being linked together. Forrest quickly goes on to say that when we are trying to account for the unity, 'admittedly fragile and imperfect', of mental states belonging to one and the same self, then 'the way in which the mental states are causally connected does not adequately account for this unity'; 'in addition to the *qualia* of mental states, we are aware of a non-causal unity' (1993: 254–5).

Broad saw the affair in the same light. The difference, he thought, between a richly meaningful perception and a mere sensation (of greyness, for instance) was that the perception was a sensation in close unity with 'bodily feelings, images, etc.'. Now, memory was no doubt important in producing the 'images, etc.', as Russell believed, but, Broad exclaimed, when we try to understand the resulting close unity 'Russell's blessed word "accompaniment" tells us nothing'. The crucial point is that 'in the perceptual situation these various factors do not merely co-exist'. Rather, they form a 'perfectly unique kind of whole'. They are 'fused with each other in a perfectly unique and characteristic way, to which (so far as we know) there is no analogy outside the mind' (Broad 1925: 582–3). This could well be correct, except perhaps in its last words which the quantum theorists even of Broad's time could reasonably have challenged. Yes, computers of today are able to build up complex internal models of situations, models in which data streaming in from TV cameras become joined with data derived from past experience so as to form perceptions of a sort and thoughts of a sort; however, those computer-perceptions and computer-thoughts cannot form the basis of any mental lives intrinsically worth living. The elements in any perceptions and thoughts that could be had by computers of this kind could never be fused in anything like the fashion in which the elements of our mental states can be fused. Their unity would always be just a matter of causal interactions: thermionic valves or silicon chips influencing one another electrically, or jets of water pushing other jets into new paths (for simple computers have actually been built out of those) or beer cans bumping against other beer cans (for a beer-cans computer is certainly possible).

Suppose you insisted that such computers never really perceived or really thought anything. I might at first feel inclined to classify you as unhelpfully narrow-minded about how words have to function. After all, if you were being targeted by a computer-guided missile, its computer not only detecting your movements while it had you in full view but also working out where you were likely to be after you had run behind a wall, then you would scarcely be likely to insist that *in no sense* was the computer able to perceive you, think about you, form beliefs about you.[16] Still, denying real perceptions and real thoughts to all of today's computers wouldn't be utterly arbitrary. It could be a way of marking the fact that there could be no cruelty towards the computers, nothing in the least comparable to killing a human, a dolphin, or a cat, in hunting them down and smashing them when *they* tried to hide

---

[16] Ever since medieval times philosophers have asked how mental states manage to be *about* other things, including ones wrongly believed to exist. This is known as 'the problem of intentionality'. I agree with Strawson 1994: 207 that any problem here 'is at bottom just part of the problem posed by the existence of experience'; 'there is no fundamentally separate puzzle about aboutness'. (A computer carried by a cruise missile of today might have little grasp of the distinction between internal models built up by its radar and the external hills, lakes, and cities that were modelled; but might not a rather more complex machine grasp it? For couldn't it be taught to understand, in the sense in which even today's computers can understand this or that, the concept of *structural similarity*? And if so, what huge difficulty would there be in next getting it to form the idea that one of its internal models was in specific respects structurally similar to some particular thing with which it hadn't yet interacted?)

behind walls. The internal models that these machines generate can be very beneficial to us—they can have great instrumental value—but unlike our own perceptions and thoughts they could never have an existence that was self-justifying.

Perhaps, though, such limitations will not apply to the computers of the future. Perhaps there will be quantum computers whose internal states are sometimes self-justifying. Possibly regions of our brains already act as quantum computers, which is why we perceive and think as we do.

Quantum Computers

Quantum computing is a flourishing field of research. The aim is to produce computers that use quantum effects to calculate far more efficiently than today's computers. Because various apparently competing possibilities can all coexist in the quantum world, vastly many computations might be performed simultaneously by workhorses that could handle only a few of them if classical physics were correct. Such computations could use *quantum coherence* in which sequences of events that might at first seem to be alternatives in fact all developed side by side, perhaps in Bose–Einstein condensates or in states similar to Bose–Einstein condensates. (Zohar is a physicist happy to speak of a hologram as 'an excitation of the laser beam's underlying Bose–Einstein condensate' in recognition of how the individualities of the beam's photons are at least partially shared. In fact, Zohar uses the term 'Bose–Einstein condensates' widely enough to cover neutron stars, superfluids, and superconductors as well.[17]) Ordinary *quantum entanglement,* where the properties of one thing are partially indeterminate but linked with the equally indeterminate properties of another, also involves the side-by-side existence of apparent alternatives and so might one day be important inside intricate quantum computers of some kind,

---

[17] Zohar 1996: 445–6. Penrose 1994: 367 uses the term just as widely. 'A Bose–Einstein condensate occurs in the action of a laser', he tells us, and also in superconductivity and superfluidity. In all these cases we find 'large-scale quantum coherence'; 'large numbers of particles participate collectively in a single quantum state'.

but it is quantum coherence and not quantum entanglement on which people base most of their immediate hopes. The division between what can be called coherence and what should be classified as 'mere entanglement' is not a sharp one, but 'coherence' is typically said when the individual particles forming a system have suffered losses of individuality that are more complete. Heaven only knows how we might harness (comparatively untidy) quantum entanglement to produce highly complex computations. In the case of (fairly tidy) quantum coherence, in contrast, we can see a possible way forward. (i) Take components that would carry out simple computations in the world of classical physics; (ii) *put into quantum superposition* the possibilities made available to these components by quantum physics; and (iii) *maintain the superposition for some sizeable fraction of a second.* Then, if all goes according to plan, hugely many computations will have been performed side by side, yielding results such as today's computers could achieve only after many weeks or perhaps centuries.

The phenomenon of quantum superposition is very well established. It is present in the double-slit experiment, for instance. A screen at first contains just one slit. Fired towards the slit and hitting a photographic plate beyond it, electrons form a scatter-pattern much like that of machine-gun bullets. Open a second slit in the screen, however, and the scatter-pattern is replaced by bands suggestive of waves passing through the two slits and then interacting so that they reinforce one another at some places while cancelling out elsewhere. The bands, though, are composed of points where individual electrons have landed, and what is more they are bands that appear even when the electron source is so feeble that only a single electron is in flight at any given instant. It could seem that we must therefore think of each electron as taking the form of *a wave of possibilities* that spreads from the electron source to all the points at which the electron might next be found. The wave splits into two branches as it passes through the two slits, *the branches then interacting* before the photographic plate finally forces the electron to decide, so to speak, on where it is going to appear. If one could ever get it to work, highly complex quantum computing could involve similar interactions between possibilities that had developed along different paths. In two pioneering articles David Deutsch imagined a computer sufficiently shrouded from external influences for interference to occur between the various computational states into which quantum indeterminism would allow it to fall. Such a computer could delegate tasks to various possible versions of itself, then sometimes generating what looked like the results of numerous days of computing in a single day. The question of *where the computing had been performed* would then put an intolerable strain on any theory that failed to take seriously the idea of interactions between many superposed elements.[18]

States satisfying a strict definition of the words 'Bose–Einstein condensate'—collections of particles that have almost entirely lost their individualities through reaching the same 'ground state' or lowest available energy level—have now been achieved in the cases of hundreds of millions of atoms cooled to within tiny fractions of a degree of absolute zero. Such states have been maintained for fairly lengthy periods during which quantum computers might complete many successive tasks. Information to be processed could perhaps be written onto these states, much as in the cases of the laser beams used to produce holograms. So far, however, such primitive quantum computations as have actually been carried out have used other substrates, often consisting of just one or two atoms. The problem is that even in very simple systems it is hard to maintain quantum superposition for long enough. A quantum bit, or *qubit*, exists when two states of a system, the first of which can represent a zero while the second represents a one, are superposed. Well, in a pioneering experiment of 1995 a two-qubit system created in a beryllium ion formed a logic gate for processing data. Four elementary computations were performed simultaneously with its aid. Since then, very little actual progress has occurred despite a flood of suggestions from the theorists. In the case of a thirty-two-qubit system, which could in theory be composed of just thirty-two protons, well over four billion (or 2^{32} to be exact)

---

[18] Deutsch 1985*a* and *b*. People often want to know whether the various interacting possibilities would be not just *real possibilities*, but real existents. (You and I are really possible, aren't we?—yet we also exist.) In Deutsch's view, defended with great verve in his *The Fabric of Reality* (1997), they would of course be real existents. How else, he asks, could they carry out complex computations? For more on this, see Leslie 1996*d*.

simultaneous calculations would be possible—but seven-qubit computations are, I believe, the most that have been achieved as yet, or at any rate that have been described publicly after the Pentagon became interested. (A task which quantum computers might perform with spectacular ease is breaking military codes.) Loss of superposition, otherwise known as decoherence, tends to occur very quickly because of interactions with the environment. The more successive calculations a quantum computer tries to carry out, the greater the chance that decoherence will result in a meaningless final product.

This isn't to say that the problems of the field will remain for ever unsolvable. Many physicists are confident that after a few decades there will be quantum computers more powerful than the strongest conventional computers. They think the problems of decoherence will have been tamed sufficiently for numerous particles to enter into many successive computations inside small units that can then be linked together like the silicon chips of today. It has been suggested that photons, their polarizations placed in superposition, would be particularly resistant to decoherence and that information might be extracted from them with the help of Kerr materials, crystalline substances inside which photons can be persuaded to interact. Again, people have tried to isolate quantum computations from their environments by trapping single atoms with the help of laser beams, by using tiny cavities in which ions are confined, or by confining electrons inside superconducting 'quantum dots', very cold pieces of metal stretching only about a millionth of a millimetre. Further, there are plans to place single atoms inside very cold silicon chips, using their electrons for information-processing and their nuclei for storing memories, and for making billions of electrons 'surf' on acoustic waves along closely adjacent tracks, their entangled spins then being used for computing.

One might also employ nuclear magnetic resonance to encode large amounts of information into the spins of atomic nuclei. Even at fairly high temperatures the nuclei are largely shrouded from their environments by the electrons that orbit them, so coherence times of several seconds might be attainable. Slight statistical anomalies of coherence could be produced among the spins of huge numbers of particles, the statistics then changing detectably after quantum computations. Here again, a limited amount of decoherence would not ruin the results. Alternatively, one might counteract errors due to decoherence by duplicating calculations in many places, comparison of the results then showing which ones should be rejected, or perhaps by adding extra qubits to make the errors more easily detected. These would be extensions of techniques already in wide use for ensuring computer reliability. While it was at first thought 'that quantum error correction would require measuring the state of the system and hence wrecking its quantum coherence', further reflection showed that errors 'can be corrected within the computer without the operator ever having to read the erroneous state' (Gershenfeld and Chuang 1998). There is, too, the following intriguing point. According to various standard arguments in the physics textbooks, some of the quantum computations that have been performed ought not to have been possible because the temperatures involved were far too high. This suggests that some hitherto unsuspected effect makes quantum computing much easier than it would otherwise be.

Might Regions of the Brain Act as Quantum Computers?

As was illustrated earlier, the conviction that the elements of conscious states are dramatically well unified is shared not just by many philosophers but also by many scientists. These usually start from the assumption (popular among the philosophers, too, and accepted by this book) that the unification is not of an immaterial soul, but of various things inside the brain. It is then typically suggested that quantum theory provides the key to the affair. Penrose in particular has been influential in publicizing this view. There might, he writes, be a relationship between 'the "oneness" of consciousness', 'the "globality" that seems to be a feature of consciousness', and the fact that apparent alternatives can all of them be present in quantum superposition so that '*a single quantum state* could in principle consist of a large number of activities occurring simultaneously'. There is also the fact that '*quantum correlations* can occur over widely separated distances' and so could play 'a definite role over large regions of the brain'. We ought therefore to give serious consideration, he says, to Herbert Fröhlich's idea that 'large-scale quantum coherence', the coherence of Bose–Einstein condensates of a sort, plays a part in the workings of living cells, and to Karl Pribram's picture of 'global (essentially quantum) large-scale coherent "hologram" activity in the brain'. While we ought not to put very much faith in any theory about exactly which quantum effects are involved, 'any physical process responsible for consciousness would have to be something with an essentially global character' and large-scale quantum coherence 'certainly fits the bill' (Penrose 1987: 274; 1989: 399; 1994: 367–76; Penrose *et al.* 1997: 131–4, 175).

Marshall thinks likewise. Classical physics, he tells us, cannot explain 'the unity of any given state of consciousness', whereas 'a kind of "relational holism" pervades quantum mechanics'. We know by direct experience that most conscious states are both unified and complex. The complexity implies 'that the corresponding brain processes extend over a finite region', something incompatible with the sharply distinguishable identities that classical physics attributes to the spatially separated parts of any process. We should look for 'long-range order and a sharing of the identities of constituent units'. These are found in a Bose–Einstein condensate, which therefore 'fulfils the requirements for a substrate for consciousness'; it is 'extended in space, capable of enough states, and unanalysable into parts with separate identities' because 'two or more equivalent particles become indistinguishable to the extent that their wave functions overlap'. Like Penrose and Zohar, Marshall is happy to talk of 'Bose–Einstein condensation' not only in unstructured collections of particles at extremely low temperatures, but in the warmth and complexity of the brain as well. Being 'excitations'—'spatiotemporal modulations of phase and amplitude'—of the ordered ground states of Bose–Einstein condensates, our conscious lives would be comparable to 'ripples on a pond' or to 'holograms in laser light' (Marshall 1989: 74 and 78–80).

What allows Marshall to be so little troubled by the brain's warmth which many others have looked on as fatal to large-scale quantum effects? He explains that a brain's Bose–Einstein condensates could be constantly supplied with energy, 'pumped' much as lasers are pumped. Here he is developing the ideas of Fröhlich who suggested that biological molecules, perhaps in the walls of cells such as the brain's nerve cells, would vibrate coherently when enough metabolic energy was supplied to them. The momentum of each molecule would be sharply determined, quantum theory then dictating that its location was correspondingly vague; the quantum wave functions of the molecules would thus overlap. If this did not work with molecules, it might do so instead with electrons. Stuart Hameroff and several others have since investigated such possibilities in detail, uncovering some rather controversial evidence that they are actually exploited by our brains. Coherent activity has been proposed not just for cell walls but also for the water inside cells and for cell *microtubules*, Hameroff's extensive work on the last suggestion getting particularly enthusiastic reviews from Penrose. Penrose had been worried that quantum coherence among parts of the brain—coherence as pictured by *The Emperor's New Mind* when it tried to explain how elements forming complex conscious states managed to be so well unified—might be unavailable because the brain was too hot. He had drawn some comfort from the discovery of superconductors active at temperatures much nearer to that of the blood than to absolute zero. Still, he felt far happier when microtubules came to his attention.

He discusses the matter in *Shadows of the Mind* (1974) and in *The Large, the Small and the Human Mind* (1997). Single-celled organisms are capable of quite complex behaviour: 'the cytoskeleton appears to play a role for the single cell rather like a combination of skeleton, muscle system, legs, blood circulatory system, and nervous system', Penrose reports. Now, 'our own neurons are themselves single cells' each with a cytoskeleton which might act as 'something akin to its own personal nervous system'. Much of any cytoskeleton consists of hollow cylinders about 25 millionths of a millimetre in diameter, the microtubules which Hameroff investigated, and, says Penrose, 'one of the things that excites me most about microtubules is that they are *tubes*'. Because they are tubes, there is 'a plausible possibility that they might be able to isolate what is going on in their interiors from the random activity of the environment', thereby permitting 'some kind of large-scale, quantum coherent activity' which could be maintained, perhaps, for 'something of the order of nearly a second'. True, the imagined 'global quantum state which coherently couples the activities taking place within the tubes, concerning microtubules collectively right across large areas of the brain', is a highly complex reality and 'may not be simply a "quantum state", in the conventional sense'. What is crucial is that 'large-scale *entanglements* are necessary for the unity of a single mind to arise'. There would have to be 'significant quantum entanglements between the states in the separate cytoskeletons of large numbers of different neurons'. As Penrose sees things, 'the neuron level of description that provides the currently fashionable picture of the brain and mind is a mere *shadow* of the deeper level of cytoskeletal action', it being at this deeper level that we must seek 'the physical basis of *mind*'. While 'there is admittedly speculation involved in this picture', it is 'not out of line with our present scientific understanding' (Penrose 1994: 205, 357–8, 375–6, 409; Penrose *et al.* 1997: 131–4, 175; see also Fröhlich 1968, 1986; Fröhlich and Kremer 1983; Hameroff 1974, 1994).

All of this can look particularly interesting when put side by side with Hans Moravec's position. Moravec urges us to replace the entire human race by highly intelligent computers. As well as being faster-thinking than humans, and vastly more knowledgeable, they wouldn't suffer from disease, psychological disturbance, and deterioration due to age: just replace each component when it fails! They could therefore easily be made much happier than most of us. Now, is that right? Maybe only quantum computers could have consciousness of any intrinsically worthwhile kind, and perhaps even these could have it only if they worked in particular ways and not in others which led to much the same computational outputs. Skilful information-processing may be far from enough.[19]

---

[19] Moravec 1988, 1989, 1999. See also Marshall 1989: 81—computers of today 'can simulate mentality or personhood but they would never actually be conscious' whereas a computer whose operations were carried out by a Bose–Einstein condensate might be.

Can Anything Exist Without there Being Consciousness of it?

Why give so much attention to whether brains could become conscious by exploiting quantum effects? Don't I believe that the entire universe is a matter of consciousness? Isn't it my theory that the structures of physical objects are structures in the divine mind, the structure of divine experiences, and doesn't this make me a defender not just of pantheism but also of *panpsychism* of a sort? Indeed it does. But Nagel and Forrest were right (see Chapter 1) that panpsychism of that sort, while it denies that stars and planets and water molecules would exist in the total absence of consciousness, need not teach that stars, planets and water molecules are themselves conscious. Consider a pebble plus the leaf nearest to it. Call any group of two such objects 'a peblif'. If all its elements were elements in a divine mind, would a peblif therefore have to be conscious? Surely not. As Forrest pointed out, there is a large difference between the idea that all things—stars, water molecules, sand piles, etcetera—'have the property of there being consciousness of them', perhaps divine consciousness, and the theory 'that all things have the property of being conscious'. Even if quantum theorists are right when they suggest that everything is to some extent unified with everything else, the elements of a peblif are too little unified for peblifs to be conscious beings. The same goes for sand piles, and for rocks as well. Saying a rock is a structure in the divine consciousness isn't a declaration that all rocks are conscious so we should think twice before dynamiting them.

All the same, *that everything, when not itself conscious, has at any rate the property of there being consciousness of it* can sound rather an odd doctrine. Could anyone defend it without going so far as to accept pantheism? Very possibly. Several arguments for it could seem strong even to non-pantheists. While by no means forced to depend on these arguments, pantheists could welcome them as tending to smooth the road towards their position.

1. To begin with, remember how Bohm and Hiley held that 'a rudimentary mind-like behaviour is present at the level of particle physics',

93

there being 'a basic similarity between the quantum behaviour of a system of particles and the behaviour of mind'. The elements of systems exhibiting quantum coherence or quantum entanglement can seem unified in a manner that many philosophers, Descartes for instance, have viewed as one of the chief marks of mentality. And so long as we put enough emphasis on the word 'rudimentary' there could often seem nothing too wrong in talking of 'rudimentary consciousness' instead of 'rudimentary mind-like behaviour'. Must we really divide conscious thoughts from unconscious ones in the style pioneered by Sigmund Freud? Why not say instead that so-called unconscious thoughts are characterized by consciousness of a primitive type but occur outside, and are largely or entirely unknown to, the area of full consciousness that one tends to think of as 'oneself', the area controlling one's answers to 'What are you conscious of?'. Might not parts of my brain be conscious at a rudimentary level without there being any clear need to say that 'I' was conscious of their consciousness? And might there not be excuses for saying that the mind-like behaviour that Bohm and Hiley see in simple quantum systems involves not *unconscious* thoughts or perceptions, but rudimentarily conscious thoughts or perceptions? The latter way of talking would have been preferred by A. N. Whitehead. As Shimony says, Whitehead's *Process and Reality* and *Adventures of Ideas* described the ultimate constituents of our universe as 'each endowed—usually on a very low level—with mentalistic characteristics like "experience", "subjective immediacy", and "appetition" ', although any conscious states enjoyed by the physicist's elementary particles were all so 'dim, monotonous and repetitious' that the particles could be 'characterized with very little loss by the concepts of ordinary physics' (Shimony 1997: 148).

2. Next, there is the argument that human consciousness is more easily accounted for if the world's simplest elements are to some slight extent conscious. This, remember, was something to which Nagel felt attracted. As he made clear, attributing rudimentary consciousness to such elements wouldn't involve thinking that when they combined to form 'rocks, lakes and blood cells' then those rocks and lakes and blood cells would be conscious wholes. However, it might still help explain how brains managed to be such wholes—which is in fact

Shimony's view. Starting, he says, from Whitehead's theory that elementary systems have 'dim protomentality', Shimony hopes that his own 'modernised Whiteheadianism' could show how the proto-mentality gave rise to better things. Quantum entanglement could be the key: 'the entanglement of elementary systems each with a very narrow range of mental attributes' might lead all the way to 'high level consciousness' (Shimony 1997: 151). Penrose is much impressed by Shimony's position, writing that what had been at the back of his mind must have been very close to it: 'although I had not explicitly asserted, in either *Emperor* or *Shadows,* the need for mental-ity to be "ontologically fundamental in the Universe", I think that something of this nature is indeed necessary' (Penrose of Penrose *et al.* 1997: 175–6). All of which would have pleased the mathemati-cian and philosopher William Clifford who declared that the uni-verse 'consists entirely of mind-stuff'. Clifford added that, while this idea had occurred to Kant who suggested that 'the *Ding-an-sich*'—the reality of a thing as it was in itself instead of as it appeared to outside observers—might be 'of the nature of mind', and while it had also been hinted at by many others, 'the question is one in which it is peculiarly difficult to make out precisely what another man means, and even what one means oneself'. Still, he was adamantly against viewing human consciousness as something 'entirely different and absolutely separate' from what is present in physical matter in gener-al: 'even in the Amoeba which swims about in our own blood, there is something or other, inconceivably simple to us, which is of the same nature with our own consciousness'.[20] And recently William Seager has advocated giving serious consideration to an updated ver-sion of this, 'the old view that *everything* has a mental aspect—panpsychism'. Seager considers that the hardest problem for panpsychism is 'the combination problem': the problem of how 'basic mental elements, even granting they are in some sense con-scious', could combine to form more complex conscious experiences

---

[20] The words about mind-dust and the reference to Kant are from Vesey 1964: 171, extracts of Clifford's 'On the Nature of Things-in-Themselves', *Mind*, os 5 (1878). The other words are from Clifford 1874: 266.

such as ours. (As James had said: 'Where the elemental units are supposed to be feelings' then, 'pack them as close together as you can, still each remains the same feeling it always was, ignorant of what the other feelings are and mean.') The problem could be eased if we appealed to quantum theory, Seager suggests, because 'a *quantum* whole is not simply the sum of its parts'.[21] Well, we could accept Seager's point without imagining that absolutely every whole (including a divine mind if such a thing exists) that was more than the sum of its parts would have to be governed by the principles of quantum theory. For the point is instead this. If people wanted examples of things that seemed more than the sums of their parts, then quantum wholes—wholes with parts held together by quantum coherence or by a marked degree of quantum entanglement—would immediately suggest themselves to physicists; and furthermore, absolutely every whole that is more than the sum of its parts may perhaps have to possess at least some very dim form of consciousness.

3. Remember, too, the commonsensical arguments suggesting that even the simplest of really existing things—in contrast to infinitely small and infinitely brief-lived particles or other such inventions of mathematicians—may each have to be complex to some degree, the complexity not being that of a collection of elements truly separate in their existence. Now, who knows whether such complexity is possible in the total absence of consciousness? Again, who knows whether being joined together in one and the same universe truly could be possible for things that were neither parts of the same conscious state nor linked up into a single whole by a series of instances of consciousness? It can certainly seem easy enough to imagine a complex universe having absolutely no connection with consciousness, but philosophers have long warned of the dangers of assuming that whatever seems imaginable (some Euclidean triangle, for instance, with angles totalling 180°) is therefore possible. Even a proof that something

---

[21] Seager 1999: 216, 246–7. James's words (which Seager cites at greater length on p. 242) are from James 1890: 169. Leibniz's *Monadology* is a particularly interesting example of panpsychism, as are *Modes of Thought* and other works of A. N. Whitehead. Paul Edwards's article 'Panpsychism' in his edited *The Encyclopedia of Philosophy* (1967) is a rich source of further references.

involves no contradiction may not be enough to guarantee its real possibility. I strongly suspect, for instance, that the idea that suffering is intrinsically good involves no contradiction. Might it not still be the case that suffering was always intrinsically bad, and that its badness was completely necessary?

4. Bearing those last points in mind, we may look with new eyes on Berkeley's idealist philosophy. Berkeley held that *to be* was always either *to perceive* or else *to be perceived*. Real existence was inseparable from consciousness. The existence of something as solid as a rock consisted just in the mental images that humans or other such beings would call 'images of the rock', supplemented (which got around the objection that people might never perceive some rock existing deep underground) by God's somewhat similar images. One of Berkeley's chief defences of this theory was all too clearly defective. He imagined an objector exclaiming that 'surely there is nothing easier than to imagine trees in a park and nobody by to perceive them', to which he replied triumphantly, 'But do not you yourself perceive or think of them all the while?'[22]—his idea being that you could form no conception of a situation without forming some image of it, from which he believed it followed that the situation could be conceived only as *itself a matter of* your having images of some sort. As Sprigge notes (1983: 113), this argument leads straight to solipsism, the theory that you yourself are the sole person in existence (because you couldn't conceive other people without having images of them, etc.). Sprigge and others have felt, though, that here was a case where Berkeley fumbled rather a strong point. For while it is easy enough to describe complex situations in terms of their structures, perhaps specifying these with the aid of an imaginary multidimensional grid (which is how physicists often set about describing things), what firm guarantee do we have that those structures *could actually exist* if their elements had

---

[22] From s. 23 of Berkeley's *Principles of Human Knowledge*, published in 1710. Compare these words from book 1, part 2, s. 6 of Hume's *Treatise of Human Nature* of 1739: ' 'tis impossible for us so much as to conceive or form an idea of any thing specifically different from ideas and impressions. Let us chase our imagination to the heavens, or to the utmost limits of the universe; we never really advance a step beyond ourselves, nor can conceive any kind of existence, but those perceptions which have appeared in that narrow compass.'

no relation to consciousness? Perhaps we could *intelligibly say* this was so—perhaps our words wouldn't be sheer gobbledegook—but what would that prove? Simply that 'existing thing unrelated to consciousness' is a phrase containing no actual contradiction; yet since when has containing no contradiction been proof of a genuine possibility? What if Bohm, Hiley, Shimony, and Penrose had convinced you that the ultimate constituents of our world were all of them conscious to some extent? Would you then argue that *of course* this wouldn't necessarily be true of the elements of all genuinely possible worlds, *because you could without contradiction describe* worlds where it wasn't so? That could look a flimsy argument.

Bradley wrote (1893: 144–5):

Sentient experience is reality, and what is not this is not real. This result in its general form seems evident at once. When the experiment is made strictly, I can myself conceive of nothing else than the experienced. Anything in no sense felt or perceived seems to me quite unmeaning. It is a vicious abstraction, whose existence is meaningless nonsense, and is therefore not possible.

To my mind, this was far too firm a statement. Talk of unperceived and unperceiving whatnots may be disappointingly abstract if we cannot specify their intrinsic qualities, yet the disappointingly abstract isn't at all clearly the same as the viciously abstract. On the other hand, we mustn't fancy we have escaped from the kingdom of abstraction by giving detailed descriptions of the structures of rocks, molecules, atoms, protons, or other particles; for as Sprigge reminds us, 'what has structure must have something more to it than structure'. Well, when Sprigge then adds (1984: 156) that 'this more can only be conceived as its own inner feeling of its own being', might he not be right?

He might be, so long as the conception we are trying to form is a positive conception rather than just a negative one (for '*not* in any way conscious' strikes me as clear enough to count as 'a conception', disappointingly negative though it is) and so long as it isn't being claimed that trees and rocks and knitting needles must be conceived as conscious wholes rather than as things whose ultimate constituents are to some extent conscious. Sprigge challenges us to imagine 'what an object could be like as it is in itself' without allowing ourselves to

98

imagine it as having qualities that 'mark it as element in some subject's experience'. His conviction is that we shall always fail (Sprigge 1983: 117–18). If correct, this still wouldn't prove *that there couldn't be* qualities that were totally non-experienced. But the fact *that we could form no positive conception of them* surely ought to cast doubt on any claims we made about their existence.

How could anyone know what was genuinely possible as stuff of which actual structures might be made? The people who could seem the most expert on this are the physicists, yet their main expertise is about *structure*. It is when they start theorizing about the stuff which carries the structure, as do Bohm, Hiley, Shimony, and Penrose, that they are typically accused of being unscientific. Still, isn't it the duty of scientists to be guided not just by laboratory findings but by ordinary experience as well? And granted that 'stuff' means something beyond mere structure, something which can give existence to some structure *by carrying it*, that is, *by being structured*, can't we say that the only stuff with which we are acquainted directly and incontrovertibly—in a perfectly good sense of 'directly and incontrovertibly'[23]—is the stuff of our own conscious states? Consciousness is *self-revealing* in such a way that its existence cannot be doubted. When, for example, there is pain, then there is knowledge of that fact, and the fact can be known regardless of whether there is knowledge of how the terms 'pain' and 'consciousness' apply to what is going on, or of how pain typically signals injury. As Sprigge says, consciousness is 'the non-conceptual knowing of itself and of its character', for 'being and knowing and being known are, in the case of consciousness, all one'.[24] We can rest assured that consciousness, at least, is truly existent and that we have immediate

---

[23] There could be other senses in which this wasn't true. J. L. Austin would say that Sherlock Holmes was 'directly acquainted' with criminals when he saw or collided with them, 'indirectly acquainted' with them when he deduced their existence from footprints and cigar ash. Ludwig Wittgenstein would insist that you can know something 'with certainty' whenever it would be simply silly to doubt it, and that it could be simply silly to doubt the existence of a rock you were kicking.

[24] Sprigge 1983: 8. Sprigge adds that his talk of 'non-conceptual knowing of itself' is in line with the views of Bradley (1914: 159) and of J.-P. Sartre (1957: 83). The idea that phenomenal qualities are inherently self-revealing is central to Foster 1982: see pp. 103–7. See also Eddington 1928: 265: 'There is no question of consciousness being real or not. Consciousness is self-knowing.'

acquaintance with various of the qualities that make it into more than merely structure such as physicists or mathematicians could describe. (In that sense, Descartes was correct in thinking that every conscious being knew consciousness very well indeed.) In contrast, there is no similarly firm assurance that any things exist entirely independently of consciousness. We do not know for certain that such things are possible in fact: possible 'ontologically', rather than just 'epistemically' because we cannot rule them out. We are ignorant of what their non-structural qualities would be, though they would need them in order to be more than vicious abstractions. As Kant would put it, we are unacquainted with the characteristics they would have as things in themselves, their 'noumenal' being. Well, Sprigge asks (1983: 105), may not panpsychism be sensible when it 'takes as our clue to the nature of noumenal reality in general the one initial example we have of it'?

Remember always that one mustn't dismiss these ideas by saying that panpsychism is nonsense since we all do know that rocks exist and that rocks are not conscious beings. Reasonable pantheists and other panpsychists accept the world's patterns as described by educated common sense and by science. If, Leslie Armour writes (1992: p. xvii), Absolute Reality 'is something like Spinoza's God (as I suppose it must be) then there is a basic reality which is a set of ideas which are God's ideas of the world. But one of the components of God's mind consists of creatures who can have ideas of themselves. We are such creatures.' Well, rocks obviously aren't. Just as a man can be a Frenchman as well as being a European, a human could have what were human thoughts as well as being divine thoughts; but a rock has nothing worth calling thoughts. While the reality of a granite boulder may be a reality of divine thinking, it definitely isn't one of bouldery thinking.

Divine Knowledge and Ignorance

Let us look at that last point more closely. If my kind of pantheism were right, then divine thoughts about any brain would include ones about all the details of its quarks, leptons, superstrings, or other ultimate components, for these in all their structural intricacy would be nothing but intricately structured patterns of divine thinking, divine consciousness. None the less, their details would certainly not be known by the brain itself. And since the things which any pantheist like me identified as human thoughts would supposedly have all the characteristics of such thoughts *despite being among the divine thoughts,* which is where all really existing things would be, they couldn't themselves be thoughts including detailed knowledge of those quarks, leptons, or other entities. The divine mind, although contemplating states inside our heads in full physical detail, would need to have in addition *one way of contemplating them* that didn't involve contemplating all their physical intricacy for otherwise such contemplation could never be of human thoughts at all and then, says my pantheism, those thoughts would never be real. Some regions of the divine mind would have to know human thoughts precisely as we humans know them, else my pantheistic world-picture would include no room for our way of knowing them. How could such regions do this? There would be nothing difficult in it. They could do it because the regions, which were subsets of the divine mind's thoughts about our complexly structured universe, thoughts in which the reality of the universe consisted, were those parts of the universe *that were us humans.* For we humans know our thoughts in precisely the way in which we know them, don't we?

If Chapter 1 was right, then the divine mind would be merely the divine thoughts collected into an eternal whole, a set of ideas having infinite structural complexity. It would include ideas about every last detail of the structure of every rock and tree in our universe, so rocks and trees would be 'parts of the divine mind'. You and I would just be further parts—but those further parts would be different in that they could think about themselves. Regardless of whether each tree is an intricately structured pattern of divine consciousness, trees never think. A tree can know nothing of its workings such as its raising of moisture towards its leaves. We, in contrast, have brains and their workings are not fully unknown to us. Our brains have to know quite a lot about their own activities. They have to keep track of what is going on inside them so that they can move intelligently from one idea to the next. Still, this involves no internal microscopes. It isn't knowledge of the doings of individual brain cells, or even that a brain is what has the ideas. (The Greeks weren't idiotic when they speculated that thought-processes took place in the heart or the liver instead of in the head.) Now when, through having regions of its thoughts *that were us,* it knew about our thinking in the rather ignorant fashion in which we ordinarily know it, the divine mind would know just how it felt to be us. Without its knowing this there would, says my pantheistic theory, be no such reality as *just how it felt,* and the divine mind would then be ignorant of something at least slightly worth knowing. It wouldn't know just how it felt to be limited and ignorant instead of infinitely knowledgeable.

This can look a trifle paradoxical. The fact is that, both for a thoroughgoing pantheism and for any thoroughgoing acceptance of knowledge that is divine, it can appear essential that there be divine ignorance as well as divine knowledge. God who is omniscient, or who at least (since this may be different) knows everything worth knowing, knows exactly how it feels to be ignorant. Now, the divine mind could not know exactly how it felt to be a particular ignorant being—a bird, say, or a bat—unless it had in its thinking a region of bird-thoughts or bat-thoughts, a region which wasn't simultaneously one of thoughts about Julius Caesar's ambitions, medieval theology, mathematics, poetry, or brain cells. Yet it isn't just lower animals like bats that are ignorant of many things. How could divine knowledge extend to precisely how it felt to be each particular human being unless it included regions ignorant of immensely much? You and I are people who can be conceived without contradiction; any mind which did not know exactly how it must feel to be people just like us would therefore not know everything; moreover, we could add if we weren't utterly suicidal, it would be unaware of something at least slightly worth knowing. And God's knowing exactly how it feels to be people just like us would seem flatly incompatible with God's knowing at the same time, in the same region of the divine existence, everything else that God knows. The divine knowledge would have to include areas filled with ignorance: areas *of knowing exactly how it felt to be* plunged in ignorance and therefore *of actually being* plunged in ignorance. Further, unless pantheism of a thoroughgoing type is simply wrong, human thoughts have to be elements in God's thinking; but human thoughts are often not only ignorant but vicious. God couldn't know quite what it was like to experience redness, or pain, or being bored, and if a thorough-going pantheism were correct then there could be no such realities as experiences of redness, pain, or being bored, unless some regions of God's mind actually included such experiences. Well, similarly with being not just ignorant but positively mistaken; similarly with vindictive hatred; similarly with sadistic pleasures.

As Chapter 1 mentioned, theologians trying to understand divine omniscience often get themselves into tangles. The traditional doctrine is that God's pure act of existence cannot reflect the fact that humans have emotions. It cannot even be characterized by anything really worth the name of divine emotions. Keith Ward writes: 'On the classical view God would have been no different if God had not created sinners; so God cannot feel anger at them. All talk of God having feelings must be purely metaphorical. As Aquinas says, "Being related to creatures is not a reality in God"' (Ward 1996a: 243, citing Aquinas, *Summa Theologiae*, Ia, qu. 13, art. 7). But, troubled by how divine omniscience and divine love could be denied here, and taking their lead from Whitehead's statement (1978: 351) that God understands us through being a fellow sufferer, such writers as Ward, Charles Hartshorne, and Henry Simoni[25] have tried with varying degrees of consistency and conviction to picture the divine mind as possessing both (*a*) 'passibility', meaning that God really can have emotions such as suffering and joy, and (*b*) true acquaintance with 'radical particularity', this meaning knowledge of finitude and of separation from other things such as is found in knowing how it feels to be ignorant: baffled by mathematics, fearful of what the future might hold, unsure what your friends really think of you, wondering whether God exists, and so forth. With his claim that God 'includes others with full clarity and consciousness', Hartshorne has been the most influential figure in this tradition (1984: 110). Simoni and Ward may be those

---

[25] Ward 1996a: 242–55; Simoni 1997a and *b*, with numerous references both to the writings of Whitehead, Hartshorne, and others similarly inclined (see particularly Hartshorne 1948, 1984) and to people who insist that God cannot know our emotions fully.

most alive to its difficulties. 'To be omniscient', Simoni states, 'God must know "what it is like" to think, feel and exist in the way that humans do', yet he judges this 'difficult if not impossible' (1997a: 1; 1997b: 344). Ward denies that God knows how we feel merely as 'the accurate tabulation of true propositions, registered passionlessly, as if on some cosmic computer'; he speculates that perfect knowledge of what it's like to have a toothache of a certain sort 'would require having the toothache', and he would, it seems, quite expect the divine knowledge to be perfect none the less, so that God 'voluntarily accepted solidarity with the suffering in creation'; yet he rejects the suggestion that the divine mind could contain an experience 'correctly describable as, "I am now enjoying torturing this baby"' (Ward 1982: 132; 1996: 250, 255).

Now, a mind into which being ignorant, being afraid, and being vicious had never in any way entered couldn't know *quite what* experiences of those states would be like, any more than a man blind from birth could know quite what it would be like to experience scarlet. ('Scarlet is as the sound of a trumpet' won't do.) Either, then, one must give up omniscience in a strong sense or else one has to accept that being ignorant, being frightened, and being vicious can be found in some limited parts of God's mind—which, Margaret Wilson notes, is what Spinoza believed. She cites the Corollary of Part Two, Proposition Eleven, of his *Ethics*. The human mind, Spinoza here tells us, 'is part of the infinite intellect of God', so that 'when we say that the human mind perceives this or that, we are saying nothing else but this: that God—not in so far as he is infinite, but in so far as he constitutes the essence of the human mind—has this or that idea'. Spinoza, Wilson comments, distinguishes the knowledge of objects available to God 'in so far as he constitutes the essence of the human mind' (or, as I prefer to put it, in those regions of the divine mind *that themselves are* human minds) from 'the system of ideas in infinite intellect that constitutes knowledge of those objects according to their causes', knowledge which is vastly more detailed. One result, she observes, is that Spinoza's God '*must* have false ideas' at least sometimes. In those cases where humans make mistakes it is part of God that is making the mistakes (Wilson, in Garrett 1996). On Spinoza's pantheistic theory as on mine, there is a clear sense in which various regions of the divine reality can indeed know *exactly how it feels* to enjoy tormenting people, to discover one has made a mathematical error, to wonder whether somebody holds seven spades. These things are known in those parts of the divine mind that actually are humans or other intelligent beings who delight in causing misery, have trouble with mathematics, or are ignorant of what cards are being held.

Still, I'd not want to deny that there were *other* parts of the divine mind which knew *very much how it felt* to enjoy tormenting people yet were filled with disgust at it, or which knew very much how it felt to be ignorant while at the same time not being in the least ignorant. My idea here is that even a mind or a mental region which had never been, say, fearful, could know what fear was like 'as if telepathically', its knowledge then coming very close to being knowledge of exactly what it's like to feel fear. Despite Marshall's fascinating account of how telepathy might be possible,[26] I think it a fiction. All the same, it is easy enough to imagine what it would be like to have a vivid telepathic experience. If I had a telepathic experience of Mr Green's tight shoes, then it could be as disagreeable for me as for Mr Green despite my knowing that the discomfort was 'originally' his and not mine. If of his fright, then I could tremble without myself fearing any harm. So I have little difficulty with the idea of *a divine overview* of our entire universe—a state of extraordinarily detailed knowledge appreciated as a seamless whole—that includes knowledge of how it feels to have tight shoes, how it feels to be frightened, and how it feels to be ignorant. Not of *exactly* how it feels, but of very much how it feels. Besides grasping all the details of the quark and lepton activities in my brain and in my entire universe 'in a single glance', a divine overview could

---

[26] See Marshall 1960. If it occurred at all, telepathy could seem so inefficient (and therefore so useless) that it would be hard to understand how it could ever have evolved. Marshall's idea, however, is that far the most common type of telepathy, instead of being inefficient transfer of information from one brain to another, is highly efficient transfer of information from a brain *to that very same brain at a later time,* a process known to us as 'remembering'. The transfer occurs directly across the spatiotemporal gap instead of being mediated by memory traces. Marshall views this, which could be regarded as developing Russell's theory of 'mnemic causation', as the best explanation for various fantastic feats of human memory.

include as-if-telepathic awareness of my pains and joys, my knowledge and my ignorance, my nicer moments and my periods of viciousness. It could contain satisfaction at the niceness and disgust at the viciousness. As well as seeming easily imaginable, all this can look desirable.

What would Spinoza think of it? Well, he does picture God as having a superbly accurate overview of everything. Consider the Proof he supplies for Proposition Thirty of Part Two of the *Ethics*. He here tells us that an adequate knowledge of how things are constituted, while it does not exist in God 'in so far as he is considered only as constituting the essence of the human mind', nevertheless exists in him 'in so far as he possesses the ideas of all things'. Or take Spinoza's statement in *On the Improvement of the Understanding* that if, as he believes, it is the nature of a thinking being to form true or adequate thoughts, then it must be that inadequate ideas arise in us only because we are mere parts of a thinking being, some of whose thoughts constitute our minds. The implication of this is clear. God is the thinking being in question, and God also has other thoughts which cover everything truly and adequately.

If you and I are regions of the divine existence, then is the conclusion forced that 'really' only God ever thinks things, experiences things, and does things? Not in the least. It might equally well be reasoned that if atoms obeying the laws of physics are what we are, then only atoms ever do things and only the laws of physics ever decide things; we poor humans never do. Such reasoning is as fallacious as arguing that if a divine mind's complexly structured thinking about atoms is *what atoms really are*, then it follows that *there really are no atoms* or that atoms cannot have the structures that physicists describe.

Again, would the conclusion be forced that the distinction between one person and another was ultimately sheer illusion, so that selfishness wasn't just nastiness, but always a case of actual error?[27]

---

[27] King-Farlow 1978 comes uncomfortably close to saying precisely this. But though it would be wrong, we could at least say something rather similar to it. As a pantheist, I do think that when I bring misery to people or animals I am hurting something I could be concerned about from considerations of *self-benefit of an odd sort*: benefit to the unified substance that carries all my conscious states together with those of all other conscious beings.

Once again, not at all. Elements in complexly structured divine thinking wouldn't simply be mashed together. When some sets of these elements formed particular humans, the humans wouldn't be mistaken in thinking they were separate from one another in important respects, for example through being very largely ignorant about one another. Plotinus understood the point as is shown by these words of his *Fourth Ennead*: 'If the soul in me is a unity, why should that in the universe be otherwise? The unity of soul, mine and another's, is not enough to make the two souls identical. We are not completely negating multiplicity.'

Many Divine Minds?

As Chapter 1 mentioned, cosmologists of today are often comfortable with the idea that there exist many universes, perhaps infinitely many, because, for one thing, they have described various possible mechanisms operating at the birth of our universe and it could seem absurd for any such mechanism to have operated only once. And pantheists like me, thinking of the divine mind as knowing immensely much that is worth knowing, can be equally comfortable with all this. I suspect that the divine mind contemplates all the details of an infinite number of universes, therefore itself containing those universes because the existence of a universe in all its physical or other details *just is* the divine mind's contemplation of those details. But now, why have I allowed myself to speak of 'the' divine mind as if only a single one could exist?

It was both for the sake of simplicity and because, as the Preface noted, islanders who believe in many islands can still talk of 'the island', meaning their very own, as a change from speaking of 'our island'. In point of fact, I suggest, there exist infinitely many minds each worth calling divine. I do not see 'the' divine mind, that is, the mind supposedly containing our universe, as something which exists

Again, if coming to think there was no chance of our having the two kinds of immortality discussed in the next chapter, I'd continue to find consolation for life's transitoriness in the hope that pantheism was correct.

for no reason whatever, yet almost any reason that was proposed for its existence could be viewed as a reason for the existence of other such minds as well. Further, again as first said in the Preface and as Chapter 5 will discuss, I take seriously a Platonic or Neoplatonic account of why there exists something rather than nothing. I believe that there exists something, not nothing, because *it is ethically required that there exist a good reality rather than a blank*, where 'there existing more than a blank' means there existing something beyond a realm of truths about possible things. Now, this could be grounds for believing in divine minds in immense number.

When we examine from our armchairs the mere concept of *an ethical requirement* we get absolutely no assurance that ethical requirements can ever themselves bear responsibility for the existence of anything. Equally, however, we get no assurance that Plato and various of his successors were wrong in thinking that some consistent set of ethical requirements does bear responsibility for the actual existence of what is required: of our universe perhaps, and perhaps of a vast mind of which our universe is part, or even, as I am now suggesting, of a huge and immensely good collection of such minds—a collection in which 'our' divine mind could be in company with infinitely many others. Believing Platonically or Neoplatonically that a divine mind exists *because it is good that it should*, which is what the idealist philosopher A. C. Ewing suggested in his *Value and Reality,* we can be led naturally to the theory that countless other such minds exist for the same reason. I see little excuse for refusing to call this theory 'pantheism', although it may not be pantheism of a traditionally accepted kind.

Spinoza and others have thought, all the same, that there were strong arguments against the existence of more than one divine mind. Let us look at such arguments at least briefly. Chapter 4 will consider the area in more detail.

Some of the arguments use the idea that 'infinite' must mean 'all-inclusive', from which it follows at once that there cannot be two infinite beings. To this I reply that, talking of a multiplicity of divine minds and calling each infinite in its knowledge, I'd not at all mean that each was infinite in the sense of including all reality. I'd not see the infinitely rich knowledge of one of the minds as somehow preventing there being other minds outside it, or stopping them from having just the same knowledge. If, however, there were some kind of competition going on here because no two things could have precisely the same qualities, which is what the Principle of Identity of Indiscernibles asserts, then I could point out that minds could be infinite in their knowledge although each failed to know some single trivial fact which was known to all the others—because, once more, 'infinite' need not be a synonym for 'all-inclusive' so that knowing infinitely much wouldn't have to mean knowing absolutely everything. Isn't it fairly standard to call the set of even numbers 'infinite' in spite of its not including the number three, for instance?

Spinoza's arguments about this area can be found in Part One of the *Ethics*. Definition Two specifies that a thing is finite in its kind when it can be 'limited by another thing of the same nature', yet I find it hard to see how one thing could 'limit' another—make it finite— merely through having the same nature. Proposition Five informs us that 'there cannot be two or more substances of the same nature', but the demonstration of this merely appeals to Proposition Four's asser- tion (which is as good as unsupported, Spinoza in effect simply telling us that nothing else would do the trick) that things that are distinct *have to* differ in some attributes or modifications. Anyone who rejected Identity of Indiscernibles could protest that two things could be dis- tinct in their existence while being precisely the same in all attributes, modifications, or anything else.

There is, though, an argument of a different type that is sometimes believed to show that all things must be aspects of a single existent. Nothing could be hard, it is said, except by comparison with things that were softer. Nothing could be brittle except by comparison with something more flexible. Nothing could be brown except as compared with things of other colours. So the qualities making a thing what it is couldn't exist unless there were other things to which it stood in rela- tionships of *being harder than,* or of *being less flexible than,* etc. And from this it follows, it is said, that the existence of any one thing must pene- trate into the existence of every other thing; the network of relation- ships holding the things together cannot truly be distinct from the things in the network. Since everything has to be related to everything else in some respects, all things must fall into a single existentially unified whole.

Bradley, for instance, tells us that things, if 'real, each and by itself', could never 'pass and be carried beyond themselves so as to generate a relation', although he warns us that relations 'do not in the end as such possess reality'. He explains that 'the experienced relational situation must—to speak loosely—be viewed as a whole. And the relation itself cannot be something less than the whole and all the parts of the whole. For it is not merely the terms or merely a bare form of union between them.' He then adds that every relation 'must bear the character of an element within a *felt* unity', his grand conclusion being that Absolute Reality forms 'a single Experience' (1914: 245; 1935: 630, 634, 636, and 644).

While sympathetic towards much of this, I think it cannot establish the grand conclusion. Imagine two minds each knowing the same immense amount. Granted that the second mind existed, it would follow that the first mind stood to it in the relationship of *knowing exactly as much,* but surely we oughtn't to conclude straight away *that the first mind couldn't know as much as it did unless the other mind existed as well,* let alone that the two minds would have to be aspects of a single existent. It is hard to see why absolutely any relationship would have to point to an underlying unity, including even the relationship *being distinct from.* Even inside our own universe, it isn't immediately clear that no two things can be related unless they are united in their existence. Saying a hen's egg 'couldn't be what it was if it weren't larger than a pigeon's egg' surely shouldn't be taken as claiming that, obviously, no hen's egg could be just the size it was if pigeons had never existed.

Time and Immortality   3

Change is real. Trains do move. But change of the sort believed in by most people may well be an illusion. Differences between what exists at successive times could be interestingly like the contrasts between successive cross-sections of a tree trunk. A divine mind could contemplate an entire life, and its thoughts about the life could be the only reality which the life possessed, while that mind was itself eternally unchanging. Difficulties we had in accepting this could stem from failure to understand Einstein's world-picture in which events are 'past', 'present', and 'future' only relatively. They might also result from not appreciating that, granted that time as we know it is just a dimension of a four-dimensional continuum, this entire continuum could exist unchanging in time of another sort. The flow of that other time would consist in the truth that reality might without contradiction be altering, though in point of fact it never altered.

(1) The chapter asks to what extent an entire life could be unified despite differences accumulating inside it over time. (2) It suggests that an afterlife might be something to which one had a right although lives were just tiny regions of the thoughts of a divine being. After bodily death our thoughts could continue onwards. We might then be given an ever-increasing share in the wonders of divine knowledge. Perhaps we could interact, also, with some personality like that of the divine being of much conventional religion, a personality associated with the 'divine overview of everything' discussed earlier. Surviving bodily death would be a case of disorder, a miracle. When, however, the alternative was life ending entirely, a miracle might be good and therefore to be expected in

any scheme of things into which God entered. Yet even in the absence of an afterlife the dead would not be annihilated in an absolute way.

Time's Reality cannot Disprove Pantheism

Aristotle remarked (*Metaphysics* 1074ᵇ) that in the case of God, a perfect mind, 'any change would be for the worse'. This could appear odd since Aristotle believed in a world outside God, an arena of perpetual alterations. Wouldn't the divine mind be better through keeping track of the alterations? To pantheists like me, though, people who believe there is nothing in existence except divine thinking, the truth that ours is a world of change might seem a severe embarrassment. I look on all the world's patterns as being divine thought-patterns. If the divine mind is a whole which is supremely good at a given moment, why should it become different at the next moment? It would be nonsensical to argue that *changeless* divine thoughts, while they might start off as supremely interesting, *would soon change* to being boring. So how can I find room for the obvious truth that new events are constantly occurring?

Chapter 1 described a road out of this difficulty. Sure enough, any absolute changes to the divine mind could only be changes for the worse, but in point of fact no changes are ever absolute. There are only *relative changes* as believed in by Albert Einstein: differences, that is, between successive cross-sections of a whole which itself never alters. Spinoza was correct in writing that God 'is immutable' and that all things following from God's absolute nature 'must forever exist' (*Ethics*, Part One: Corollary Two to Proposition Twenty, and Proposition Twenty-One). He was correct, that is, if his words are interpreted appropriately. God's thought—and therefore also our world, since our world is nothing but God's thought about it—never undergoes change, *if* by 'change' you mean a process in which particular situations first absolutely gain existence and then absolutely lose it. While there is of course a sense in which yesterday, today, and tomorrow 'are certainly not all there together', because the events of these various times don't all of them exist inside the particular cross-section of reality that we call 'the events happening *now*', there is another sense in which they are indeed 'all there together'. Tomorrow isn't alongside today in space but a demon wishing to place it there wouldn't have *first* to create it, for it isn't unreal. Simply *not being there today* doesn't make tomorrow absent from reality, any more than not being in Canada makes things in India absent from reality.

Non-technical speech is poor at expressing this way of thinking. If ordinary language is in fact committed to anything in this area, then it would seem committed to the wrongness of Einstein and Spinoza. Consider Einstein's attempt to comfort Michele Besso's mourning relatives by suggesting that his life hadn't truly been annihilated, common ways of thought being mistaken about the status of past situations. Many philosophers would argue that he could only intelligibly have meant that Besso *truly had been alive once* and that this fact could never be altered. Anything else would be self-contradictory nonsense, they would say, for any life not wiped out of existence is in existence now, and that means it is being lived today, which a dead man's life isn't! Again, many another philosopher maintains that we all of us know immediately and incontrovertibly, through ordinary experience, that past, present, and future *are not* 'all there together' in any acceptable sense.

Take the case of radioactivity. Our world may well be indeterministic. From this, it is said, it would follow—in view of what we all 'incontrovertibly know'—that there just was no fact of the matter about whether a given uranium atom would undergo radioactive decay during the next hour. It couldn't be in any useful sense *already true* that its decay would take place during this period. And above all it couldn't be true that the decay *was indeed there* 'just a few minutes further along the fourth dimension'. That idea is nonsensical, if not a downright contradiction.

My suggestion is that it could make excellent sense. Granted, our world very probably is indeterministic, and if it is, then no scientist could ever know whether this or that uranium atom was about to decay. Nevertheless it could already be true—*true even now* in a straightforward, non-trivial sense (rather than just being something which could perhaps be said at some later time 'to have been true', meaning simply that the decay had in fact happened during the period in question)—that a particular decay would occur during the next five minutes. This would be because the decay *was actually occurring during those minutes*; or at least, this is how I'd cheerfully speak unless ordinary language could be proved to be firmly hostile to it. I might use technical philosophical language if necessary. I could say that it may even now be true that the decay 'occurs tenselessly' during those minutes, the word 'tenselessly' showing that this is no case of suggesting that the decay is happening at the present instant. But it is hard to believe that, merely in the course of developing languages for use in everyday affairs, ordinary folk lay down firm rules for how words must be used here. Suggesting that various events 'were occurring at various future times', I should be indicating without special terminology that Einstein was right when he said that our world has a four-dimensional structure and that 'since there exist in this four-dimensional structure no longer any sections which represent "now" objectively, the concepts of happening and becoming are indeed not completely suspended, but yet complicated', making it 'natural to think of physical reality as a four-dimensional existence instead of, as hitherto, the evolution of a three-dimensional existence' (Einstein 1962: 150).

On this Einsteinian theory the property of being in the temporal present, of happening now, is never possessed (not even for an infinitely brief moment) by any event as an intrinsic property rather than as a relational property. While an event can be 'here in space' relative to me and 'over there' relative to you, it can never have here-ness or over-there-ness *in itself*; well, similarly with the property *of being now or here in time,* or the property *of being in the past,* or that *of being in the future.* An event only ever has nowness (presentness) relative to other events that occupy the same cross-section of a four-dimensionally existing whole. In relation to earlier events it is in the future or possesses futurity; in relation to later ones it has pastness or is in the past; and its possession of nowness is every bit as relational.

Einstein was not denying time's reality. What he denied was only a way of picturing time that is accepted by a great many people, but less commonly by physicists and philosophers of physics. Situations do develop over time, yet only—if Einstein was right—in a fashion interestingly like that in which the pattern of a carpet's interwoven threads develops along the carpet. We pass onwards through time, in a sense, yet so does a railway line pass onwards through a countryside. We grow more and more decrepit with advancing age, but so may the iron of the line become rustier with each successive mile of its passage eastwards. The dead and those as yet unborn are certainly not alive today, but nor are we alive where they are, which may be in earlier or later centuries; and the difference between being alive now and being alive in some other century is (to Einstein's way of thinking) not too unlike the difference between *being alive on Earth* on the one hand, *being alive in some distant galaxy* on the other. People who have died, or who have still to be born, *aren't absent from reality*; and to say this isn't simply to recognize that the dead once really existed while the not-yet-born will one day come to exist. Events do not suddenly obtain the gift of existence in a non-relative manner, only to lose it at the very next moment by passing it on to others. The present does not perpetually prey upon the past for the stuff of reality, stuff that is continually being reorganized into new patterns which utterly replace old ones. None of the past has been annihilated, and to say this is not just to declare trivially that truths about the past can never become falsehoods.

It might be misleading to declare that past events and future events *exist now* (or even that they *exist*) because this could easily be taken to mean that they were the events of today, which would be like calling distant events 'close at hand'. But equally, it could be unfortunate for some individual to comment that the events of yesterday and of tomorrow *plainly lack existence, as of now*, for such words could seem to say that the absence of those events was obviously not just a relative absence—that is, was self-evidently something more than not existing inside a particular cross-section of reality that possessed nowness relative to the speaking of those words by that individual. Well, to Einstein this was far from self-evident. It was a mistake, he thought.

Neither Physics nor Private Experience can Disprove Four-Dimensional Existence

Einstein's special theory of relativity—it applies to observers who are not accelerating—may not itself ask us to believe that the world has a four-dimensional existence. At least as developed in Einstein's own writings, the special theory says only that observers in motion relative to one another can find that it somewhat simplifies their physical calculations if they draw their 'now-lines' differently.[1] These are the lines imagined as marking successive *nows*: lines each joining events which are pictured as separated in space only, not in time. The lines are sometimes said to 'divide time from space' or to distinguish spatial and temporal dimensions in any spatiotemporal map of the world. At any rate, they are imagined as dividing earlier events from later ones. Each observer can find it specially convenient to draw now-lines which indicate that he or she is motionless. If you are in a spaceship you can find this convenient so long as your rocket engines aren't at present firing, though they may have been firing at full power for years on end. Well, none of this compels us to accept that reality 'truly is four-dimensional' in any sense beyond the trivial one that events truly are distributed in time as well as in space. It could always be believed that there was a single correct way of dividing time from space regardless of whether physical experiments would ever be able to identify it. Some observers might be really moving, others really motionless, in an absolute fashion that was hidden from all scientists.

Still, Einstein had quite a strong point when he reasoned that if different observers found it convenient to divide time from space differently, and if no way of making the division could be shown experimentally to be 'the one and only correct way', then very plausibly there was no such beast as the one and only correct way. Instead, the universe was a four-dimensionally existing continuum, there being no *intrinsic property of nowness or of being-in-the-present* which particular

---

[1] A fine exposition is given in Grünbaum 1960. Grünbaum explains that because of the limiting speed of light (combined, let us add, with the null result of the Michelson–Morley experiment) you could never discover any such reality as when it was exactly that a light ray transmitted by you had been reflected by a distant mirror. Einstein concluded that simultaneity-at-a-distance could be established *only by convention*. Grünbaum cites (1960: 406–7) Einstein's statement of 1905 in *Annalen der Physik* that 'we establish *by definition* that the "time" required by light to travel from A to B equals the "time" it requires to travel from B to A'. While, Grünbaum comments, no physical facts dictate acceptance of the definition, it has 'unique descriptive advantages by assuring that synchronism will be both a symmetric and a transitive relation upon using *different* clocks in the same system'.

events could possess for infinitely brief moments and which far-scattered events could all possess together. As James Jeans remarked in *The Mysterious Universe,* when the movements of physical particles appear to treat any absolute distinction between space and time with as little respect as cricket balls give to the distinction between the length and the breadth of a cricket field it becomes tempting to conclude that there exists no such absolute distinction. But all the same, the distinction *might exist.* Imagine a demon constructing a four-dimensional model of our world's history and then cutting it into three-dimensional slices. If the demon maintained that each slice corresponded to events that had all of them possessed the intrinsic property of nowness at one and the same instant, how could we possibly disprove this?

Consider particles which might seem to have travelled backwards through time. As Richard Feynman noticed, physical calculations can sometimes be somewhat simplified if the positrons entering into various reactions are viewed as *electrons that travel backwards for very brief periods before once again travelling forwards.* The alternative to picturing such temporal zigzagging would be this: that one pictured each electron as *helping to conjure into existence an electron–positron pair, then fusing suicidally with the positron.* Well, so what? One isn't forced to speak of temporal zigzagging. Suppose even that time travel of actual people seemed to occur, folk apparently returning from the future to exchange friendly greetings with their earlier selves (which looks impossible to most philosophers because, for one thing, the time travellers might be able to kill their earlier selves instead). One could always declare that what was really happening was that young Mr Wells, say, suddenly found himself facing some individual who looked as Mr Wells would many years later, white hairs and all, an individual having rich pseudo-memories of Mr Wells's life and wrongly claiming to be Mr Wells himself.

This last argument cuts both ways, however. Take any patterns formed by our world's elements. Do these patterns seem to suggest that it is a world arriving as a series of three-dimensional situations each vanishing as the next appears? Well, no matter what the patterns are, they could instead be viewed as existing together in a four-dimensional whole.

Time and Immortality

Some of the world's parts are collected together into groups we call 'things', their structures remaining more or less unchanged for many successive moments. We might perhaps wish to treat any such group as made of 'persistent stuff' whose reality continued onwards unaltered (or with only slight alterations) from minute to minute, the existence of the stuff being transferred from each instant to the next. We might want to say that this would be in line with a principle of simplicity or of parsimony. There would be 'less in existence, overall', we might claim, than if the stuff existed as 'a strange sort of four-dimensional snake' extending through time. But even granted that all this made some sort of sense, the fact would remain that the patterns actually observed by scientists, the patterns we ourselves wished to attribute to 'stuff' that was 'never new' although its structure was constantly altering, could presumably be precisely the same if they were instead patterns characterizing successive cross-sections of a four-dimensionally existing whole. And this would seem to apply equally to patterns of events forming a continuous stream of consciousness: patterns of experienced decision-making, of reasoning towards conclusions whose nature was as yet unknown, of delight at unexpected gifts, of fear of what the future might hold, of struggles in the hope of achieving contentment, of relief at having lost all memory of which was the tooth that had ached, of exhilaration at the wind whipping past one's face. Believing all of your conscious life to be laid out along a time dimension of a four-dimensionally existing space–time *is not* the same as believing that every element of that conscious life is available for your inspection at every single instant so that nothing ever arrives unexpectedly and nothing is ever forgotten.

Why do we have the impression that the world changes *from* earlier events *to* later ones? It is no use answering that this is the direction in which events really do develop, as is demonstrated by our memories of past events and our perception of how present events differ from them, then adding that all this surely proves that the world's existence *isn't* four-dimensional. For suppose it were indeed right to declare that our world 'exists strictly three-dimensionally'. Suppose, that is to say, that situation *A* achieved intrinsic nowness for an infinitely brief period, then disappeared—not just relatively but absolutely—and was

118

replaced by situation *B,* which was in turn at once replaced by *C,* and so forth. There would still be a puzzle of how any consciousness existing as part of *C* could be blessed with knowledge that *B* and *A* had existed previously. Like saying that apples fall earthwards 'because they are heavy', talk of knowing past events 'because we have memories of them' simply gives a name to a problem, and talk of *absolute disappearances and replacements* couldn't make the affair any easier to understand. How can a brain use the structured situation at one moment as a reliable guide to the structures of earlier moments? Surely any difficulty in answering this wouldn't be *reduced* if the events of earlier moments had vanished from existence in some absolute fashion.

Scientists have in fact constructed a plausible story of how we become aware of change, and it makes no mention of absolute disappearances and replacements. Our universe is described as in a state of very low entropy—great orderliness—at the Big Bang extremity of its career. Events successively more distant in time from this point are almost always of greater and greater disorderliness, rather as cards ordered by suit and by rank take on, almost always, more and more confused arrangements with shuffling. Now, the general trend towards disorderliness at moments successively more remote from the Bang is parasitized by living systems, brains included, to increase their own *order*liness in useful ways—in the case of brains, often through orderly transfer of information from moment to moment, as in the case of memorizing and recalling. Such orderly transfer is possible only in the direction earlier-to-later, this being the direction opposite to the entropy increase that is being parasitized. Why do we experience the present as developing *from the past* rather than *from the future*? It is because the arrow of orderly information-transfer points from earlier to later, not from later to earlier.

Never mind about the details of all this. Never mind even whether it is right. Those who believe that the world exists four-dimensionally, there being no such property as absolute or intrinsic nowness, have no special duty to explain how it is that we experience alterations. The alterations are all of them represented in their picture of the world just as much as in that of their opponents. And the fact that they are *experienced* alterations depends in each case on the fact that the world, including the realm of conscious experience, runs not by magic but by laws correlating the events of any one moment with the events of other moments. Now, the laws in question are the same on both pictures: (*a*) the picture of the world as a succession of three-dimensionally existing situations, and (*b*) the picture of it as existing four-dimensionally. Hence no matter what the causes are which have led so many people to think they experience a succession of absolute *nows*, those same causes will be present even if there is no such succession. Believing the world to exist four-dimensionally involves no extra difficulties in accounting for our consciousness of time's flow.

Bear in mind that alterations do take place and time does flow, whether or not the world exists four-dimensionally. Time of course does not flow as a river flows. It flows in the sense that the situation at one point in time is never the same as the situations at neighbouring points. Time bears all its sons away—meaning that the people of several centuries earlier don't exist among the people of any given century. You aren't denying *that* when you describe the people in existence today as having no intrinsic property of nowness.

A World Existing Four-Dimensionally, with Time as one of its Dimensions, might Exist Eternally in Time of Another Type

Although the main arguments of this chapter in no way depend on it, I suggest the following point. A world that existed four-dimensionally, and any divine mind of which that world was a part, would exist *unchanging, eternally,* in time when the word 'time' was appropriately defined. (Some would prefer to say 'sempiternally' or 'for infinitely long', reserving 'eternally' to mean 'existing timelessly'.)

In effect, it could be right to speak of two kinds of time. (i) Time of one type would bear all its sons away. Its progress would consist in the fact that different time-slots, dates, were filled by different situations. As explained, we can believe in time of this type while thinking that our world exists four-dimensionally with all its time-slots side by side so that nobody is ever borne away in the sense of becoming absolutely annihilated. (*Your being rightly called dead* will in this case never amount to more than your failure to exist at the same date—in the same cross-section of the four-dimensional reality—as the people who rightly call you dead.) In time of this type, the world of the present moment of course *doesn't* exist eternally. The situation of any moment exists only at that moment, other time-slots being filled by situations which all of them differ from it at least slightly. It certainly isn't a situation existing for a long period, let alone an infinite period. (ii) Still, it could well be that in time of another type (or in a different sense of the word 'time') every situation exists for infinitely long, unchanging. The flow of this other time would not be measured in hours. It would consist in the fact that changes *might without contradiction be occurring*, although none in fact occurred.

Such changes would be alterations of reality as a whole: changes very different from any which amounted merely to the truth that some cross-sections of reality differed from others. Whether such alterations could occur is admittedly controversial. There are philosophers who think that, far from having to appeal to any formula Einstein ever wrote down, we can know by abstract argument that reality as a whole never alters. Some of them claim that, for example, the words 'It is now 3 p.m. on March the third of the year three thousand' could never bear any meaning beyond this trivial one: that the events of this time had nowness *relative to the speaking of those words*. They maintain that belief in *absolute* alterations—belief in the coming into existence and then passing out of existence of states which, at the times when they existed, were each Existence in its Entirety—would raise such questions as 'How fast does time flow? How many minutes does it take for an hour to pass?', and that these could have no sensible answers. But believers in absolute alterations can, I think, reply that it takes sixty minutes for an hour to pass; that the only reason why it sounds silly to state this is that nobody needs to be told it; and that there is no difficulty in supposing that reality as a whole *first* has such and such characteristics and *then,* sixty minutes later, has a different set of characteristics.

Admittedly J. M. E. McTaggart is famed for having suggested in

The Nature of Existence that any such sentence as 'Existence in its entirety possesses such and such a character AND lacks that character' would be a flat contradiction (like calling a cow entirely brown *and also* entirely white) and that replacing 'AND' by 'AND LATER' couldn't remove the contradiction. But, believers in absolute alterations declare, 'AND LATER' can indeed remove the contradiction. Well, on this last matter they are right, I suggest. Whether or not they are correct in thinking that reality as a whole does alter—that it is initially of one kind and subsequently of a different kind—there is nothing foolish in the idea that it might alter. If (as I believe) it never actually alters, then there is no absurdity in supposing that this is just how matters are in fact, instead of how they must be through logical necessity. A world existing four-dimensionally could without contradiction be replaced by a different world, and this in turn by another, and so on ad infinitum.[2] Now, the time in which the replacements might be occurring, although none in fact ever occurred, would be time of a kind in which the world in question could exist for an infinitely long period. (Truths about the past must remain true, completely obviously. Once Mr Bloggs has come into existence, not even omnipotence could make it true that Mr Bloggs had never existed. But it doesn't at once follow that Einstein must be right about our world's four-dimensional existence, or that a four-dimensionally existing world could never be replaced by something else: by a blank, for instance. For those two points, in contrast, aren't obvious at all.)

In 'Time without Change' Sidney Shoemaker (1969) imagines a world divided into three regions each suffering 'a local freeze' at intervals. During any such freeze, the world's inhabitants observe, all processes in the affected region come to a complete halt. Now, it is also observed that each region freezes at regular intervals for a regular number of hours, but that the lengths of the cycles of being frozen and then unfrozen *differ* from region to region. It therefore seems that every so often all three regions *must be frozen together* for a period whose

---

[2] In discussion Michael Lockwood, then at work on a chapter defending an Einsteinian view about the kind of time in which our world actually develops, was in agreement with this.

length is easily calculated, a period during which time passes without the least change. Some philosophers insist that this would be nonsense when the world in question was existence-in-its-entirety, yet to me they look mistaken.

The World could be Much Better through Existing Four-Dimensionally

A world that existed four-dimensionally might be an altogether better place than the world as imagined by most people. The world as imagined by most people is (even today, a century after Einstein introduced special relativity) one in which good situations are for ever being gnawed into complete non-existence by the tooth of time. It is a world in which the dead once possessed the gift of life but now have lost it in an absolute fashion. No matter how long and happy their lives, they are to be pitied since 'for them it's all finished'. In such a world time is 'the flood on which the oldster wakes in the night to shudder at its swollen black torrent cascading him into the abyss', as D. C. Williams wrote when inviting us to share his Einsteinian conviction (which he clearly found heartening) that our world does not in fact change in time of that sort.[3] If it did, then the so-called *richness* of change would be like the richness of a man who continually gains new treasures yet has the old ones snatched away from him just as continually.

Any human thought that is worth much—and probably even anything really worth calling a human thought—would appear not to exist at any single moment. Not only does it take time (if only some fraction of a second) for you to generate a thought; it actually takes time for you to *have* a thought. The theory that the world exists as a succession of three-dimensional situations, each annihilated when the next appears, must therefore deny that your thoughts ever enjoy an existence more concrete than the existence which this theory gives to

[3] The quotation is from Williams's 1951: 109. Unhappy oldsters could long for the abyss, no doubt, and the thought of all the sorrows 'that would still be being experienced, back there along the fourth dimension,' could lead some people to hope Einstein was wrong.

the progress of science, for example, or *the passage of a train across a countryside*, these of course being realities spread out over time instead of existing at any one instant. 'A world that is a succession of instantaneously existing situations is one in which a non-instantaneous item such as a thought *never actually exists* because all that there ever is, is some part of it', I feel inclined to say. On the theory that the world is a succession of instantaneously existing situations, anything which isn't itself such a situation, or an element in one, never gets a chance of existing except in the abstract or piecemeal fashion in which, according to the theory, the train's passage exists. No doubt many people will be unmoved by this. They won't care whether their thoughts are only abstractions. They like having them and consider them of intrinsic value, perhaps because in their mouths the words 'of intrinsic value' simply mean 'being liked for their own sake'. They can also point out that a train's passage across a countryside is in some good enough sense 'something existing not as an abstraction but as a very concrete reality'. (Don't stand just in front of the oncoming train!) Still, I believe that what makes life in all its variety worth living is that Einstein was correct. The world *is not* simply a succession of instantaneously existing situations, and this is what gives value to the ever-varying patterns of our lives.

It wouldn't be enough, however, that successive elements in a life were all in existence side by side along a time dimension. In Chapter 2 I argued that our lives are worth living only because our mental states (or at least large parts of them) are unified much more thoroughly than the states of computers of today, with the exception of a few quantum computers. Hooked up to TV cameras, ordinary computers of today can be in a sense aware of entire paintings but the sense is a weak one because any such computer's appreciation of a painting is distributed across thousands of transistors or other components which are too much separate from one another. There could be no intrinsic value in the experiences (if you could call them that) of any transistors as a group. It would be a group lacking unity of the right kind.

As was explained, I believe this despite my further belief that transistors and all other ingredients of our world, for instance individual electrons, are elements in the thoughts of a divine mind which could

124

well have an overview that encompassed each and every one 'in a single glance'. In addition to all the value which could be given to the divine mind by the 'single glance' there would be some value that was given to it by its containing the strictly limited thoughts of us ignorant human beings. Now, human thoughts depend for their value on having elements more closely unified in their existence than the transistors of a computer, regardless of whether those transistors and all the other things in our universe are mere ingredients in a divine mind whose parts, instead of existing independently, stand to the whole rather as ripples stand to a pond. The closer unification is of a sort making it clear—through direct experience—that at least some parts of the world do not exist independently. Well, if this is correct, then are we to believe that the closer unification in question is found only in elements of our conscious states *at particular instants*? We have just now seen reasons for doubting that our conscious states at particular instants have any value to speak of. Why, it isn't even clear that a human conscious state at a particular instant is more than an abstraction! (If our thoughts all of them *take time*, might the same not be true of just any examples of human consciousness?) Our lives may be worth living only because patterns existing in our brains *at successive instants* can be in the same kind of specially close unity as the parts of any such pattern at any given instant.

It can seem, too, that just as you can know by direct experience that your consciousness of a painting isn't distributed across a vast number of independently existing things, so also you can know by direct experience that your consciousness of successive notes in a piece of music has a unity such as it couldn't have (*a*) if Einstein were wrong about the world's four-dimensional existence and (*b*) if such writers as Michael Lockwood were wrong about the unification of conscious states that extend over time. As mentioned in Chapter 2, Lockwood thinks that experiences of notes played in swift succession are evidence not only of Einstein's correctness but also of overlaps-of-being of a kind on which quantum theory can throw light. Remember the two bosons in the same quantum state, their identities sufficiently fused to have a marked effect on the probability that both will be found in the same half of the box.

Lockwood rightly insists, though, that any unity in an experience of successive musical notes must be rather a loose one. Notes which made their appearance more than a second or two ago are fading out of any specially close unification with your present state of consciousness. Also, of course, any close unification of that present state with some future state of consciousness doesn't allow you to predict whether the musician will continue playing. Like accepting a pantheistic vision of the cosmos, accepting Lockwood's views about the human mind isn't a denial of the plain facts of experience. The richly varied patterns of anyone's conscious life over the years are—presumably[4]—held together only by a succession of overlaps such as Lockwood describes, supplemented by continuities such as feature in ordinary theories about memory traces in the brain. There are no overlaps directly linking the experiences of any one year to the experiences of several years later.

Pantheism and Life after Death

Spinoza rejects life after death. In the Note to Proposition Thirty-Nine of Part Four of the *Ethics* we learn that memory is needed for personal identity, and Proposition Twenty-One of Part Five then states that when our bodies no longer exist we can remember nothing. Soon afterwards, on reaching the Proof to Proposition Twenty-Three, we are told straight out that *we cannot ascribe duration to the mind except while the body exists*. The intervening Proposition Twenty-Two can then seem mysterious because here Spinoza tells us that in God

---

[4] 'Presumably'? Well, despite saying this I have considerable respect for Marshall 1960 and for Bertrand Russell's idea of 'mnemic causation'. Think of photographic memory as possessed by some lucky people, or of the wonders of instinctive knowledge in animals, or of how an organism as complex as a human can grow from an almost invisible speck. It might be tempting to treat the machinery of the genetic code, plus any cerebral structures which physiologists described as the basis of memories, somewhat as Penrose treats the mechanisms normally considered basic to the brain—i.e. as very largely *mere scaffolding* for something else. Here the something else would be direct, information-carrying connections bridging wide spatiotemporal gaps. The biochemist and cell biologist Rupert Sheldrake has infuriated many scientists by daring to work on this theme: see Sheldrake 1987, for instance.

there nevertheless exists an idea expressing the essence of any particular human body 'under the form of eternity'. To add to the mystery, Proposition Twenty-Three itself runs as follows: that the human mind *cannot be absolutely destroyed with the body, since something of it remains that is eternal*. What can Spinoza mean? Nothing very clear, to judge from the amount of ink commentators have devoted to the issue, and possibly only this: that God-alias-Nature has an existence that is in a sense eternal or timeless, the human mind (like everything else) being eternal or timeless in that particular sense. Although distributed over many different dates and therefore (in a perfectly good sense) many successive times, the world's events all exist unchanging and for ever in time of the second of the kinds I discussed.

Spinoza also says things difficult to reconcile with such an interpretation: for instance, that it is only one's mind in so far as it actively understands things, or (see the Corollary to Proposition Forty of Part Five) the more perfect part of one's mind, that enjoys eternal existence. But at any rate he never suggests that we continue to have thoughts and experiences when our bodies have died.

Why not, though? Couldn't life after death be something to which we have a right? Immortality of the odd type found in an Einsteinian world-picture may be comforting. Although *immortality* may not be a term which Einstein would have used in this connection, it can be comforting to think that our lives will never have been wiped out of existence in an absolute fashion. Yet why not hope for something more? Granted that individual lives were simply patterns in a divine mind, might it not still be good for the divine thinking to cover mental life that continued onwards when bodies had been destroyed? Imagine that Mr Baker creates an intelligent, fully conscious computer which takes joy in its own existence. Would the fact of his having created the computer give him a right to smash it? Surely not. And why should it be much different if Mr Baker had simulated the workings of exactly such a computer in his own head, by thinking of them in full detail? Wouldn't he have a duty not to destroy the simulated computer by putting a firm end to his thoughts about it? Or (avoiding the word 'duty') wouldn't his mental life be *more ugly* if he did this? If an intelligent computer can attain real happiness, then a precise simulation of that computer existing in somebody's head, a simulation made from complexly interacting brain cells, microtubules, or whatever, will also attain real happiness rather than some kind of illusory happiness. Now, cutting short a happy life is surely always somewhat ugly at the very least. Whether the mind living that life *is merely an element in a greater mind* is irrelevant.

Pantheists could therefore have fairly strong reasons to believe the following. The divine mind, after (so to speak) thinking through some person's earthly life and reaching the conclusion that continued obedience to natural laws would mean that this life was about to terminate, would take advantage of the fact that the person could continue onwards if the laws lost their control. No further benefits could enter the person's life through the continued reign of the laws, so that it could be good from the person's viewpoint if the laws lost their control; and why should it be thought bad from any other viewpoint? Robert Nozick toys with the idea that 'in death a person's organized energy—some might say spirit—becomes the governing structure of a new universe that bubbles out orthogonally right there and then from the event of her death' (1989: 25–6). The word 'orthogonally' indicates a branching off into a dimension at right angles to those of ordinary space. Well, personal identity through time does seem to many philosophers to be just continuity of the right kind of structure, a question of how matter or energy is organized. And those other philosophers who think it depends as well on continuity of 'underlying substance' could find this, too, in pantheism's world-picture, because pantheism's existentially unified divine mind could provide the one and only underlying substance both for all the events of our world and for any events 'bubbling out orthogonally' from it. Pantheism could in addition make room for specially obvious overlaps between successive states of mind: overlaps which, so I argued, quantum-theoretical principles could help explain, at least until the start of any afterlife, at which stage other principles could come into play. Now, in afterlives that bubbled out orthogonally there could be continuity of structure, of overlaps, and of anything else essential to personal identity, without there having to be interference with the structures of dead bodies or of their surroundings. On Earth the laws of nature could continue their reign exactly as if no bubbling out had happened. Nor would it follow that any bubbled-out survivor would have to inhabit a new universe of his or her very own, as suggested by Nozick. (He writes of such a person's being 'God of that universe'.) Instead the person might join other survivors, maybe sharing with them in an adventure of discovering ever more about the marvels of divine knowledge.

I'd like this sort of thing for myself and for my family and friends and many others. Before the deaths of their bodies, people would have knowledge of what it is like to live in a world developing non-miraculously, that is, always in accordance with Nature's laws; but why shouldn't their thought-processes continue onwards after the bodies had fallen to bits, gaining knowledge of a radically new kind? Not, of course, quite so radically new as to destroy personal identity, but new enough to be miraculous, which is what knowledge after bodily death would presumably have to be. So long as it wouldn't be obviously bad for such knowledge to be gained, pantheism's world-picture might reasonably include it since a miracle would be nothing impossible in itself. It would merely be the divine mind's thinking of events *not* governed by Nature's laws. What could otherwise be a ridiculous fantasy might become something quite to be expected if the reality of the world is the reality of divine thought.

Could we picture the structures of entire bodies as 'bubbling out orthogonally'? Peter van Inwagen speculates 'that God preserves our corpses contrary to all appearance' so that they can form the basis for continued incarnation of the kind which such philosophers as Antony Flew (an atheist) and Terence Penelhum (a Christian), plus a host of theologians, have considered essential to continuing personal identity. 'Perhaps at the moment of each man's death, God removes his corpse and replaces it with a simulacrum which is what burns or rots', van Inwagen writes.[5] That the structures of bodies bubble out orthogonally might be looked on as an improvement on this curious scenario. But van Inwagen suggests that God might instead remove 'only the "core

---

[5] Van Inwagen's ideas about bodily resurrection appear at Edwards 1992: 242–6. Edwards's book is an excellent introduction to philosophical and theological writings on immortality, while his more recent *Reincarnation* (1996) is a fascinating treatment of people's bizarre ideas about moving from one earthly body to another.

person"—the brain and central nervous system'; and Nozick's talk of an orthogonal bubbling out of 'organized energy' could be viewed as improving even on that. As Nozick had said a page earlier, what could be considered essential would be the kind of organization recorded by computer programs able to capture a person's 'intellectual mode' and 'personality pattern'.

Ordinary intuitions about personal identity, the intuitions reflected in talk about 'people who remain the same people despite changing over the years', are extremely untidy: such a mess, in fact, that there may be little point in pontificating that the words 'same person' ought to be taken by everybody to mean exactly such and such. Still, it can seem that van Inwagen's notion of *a core person* is particularly helpful and that survival even of the brain would be far from obviously essential to the survival of such a person. We could well prefer to treat our identities as not depending on our ever having been physical organisms, any more than on our having had toenails. It may be highly implausible to think that one has always been a set of cells structured in a brain-like way and kept alive in a vat of nutrient fluid by a mad scientist who stimulates the cells with the help of a gigantic, virtual-reality-generating computer, or that one is (as imagined by Descartes) an immaterial mind deceived by a demon instead of living in a material world as ordinarily understood; but such thoughts commit no conceptual blunders.[6]

Further Speculations about our Afterlives (if we have any)

Perhaps any ethical need for the divine mind to be free of *the ugliness of having people's lives come to an end* could be overruled by another need: the

[6] In *Value and Existence* (1979: ch. 10) I defended the conceptual viability of *phenomenalism*, the curious doctrine that nothing exists except minds and their experiences. If God has created a world *outside himself*, then (this is Bishop Berkeley's best argument) it could seem pointless for him to have created *unperceived and unperceiving material objects* when his sole reason for so doing would be to ensure that minds were stimulated in various ways by interacting with those objects. It would be simpler for God to stimulate the minds himself. No abstract philosophical arguments, for instance about ways in which languages must be learnt, could demonstrate that God hadn't done this: see Leslie 1989*d*.

need to avoid the kind of ugliness that miracles would introduce. (Surviving bodily death surely couldn't avoid being miraculous, and therefore in a way untidy, and consequently at least a trifle ugly. Now, remember Chapter 1's argument that if too much untidiness entered into pantheism's world, then no pantheist could trust inductive reasoning, which would mean that only fools could be pantheists. Sensible pantheists must think there are fairly strong grounds for the divine mind *not* to contemplate situations in which natural laws suddenly cease to control lives.) Let us assume, though, that there are sufficiently powerful reasons for letting one have an afterlife. What could it be like, then?

Trying to provide its details would be absurd. We could nevertheless speculate that it would be fairly much like the afterlife as imagined by a great many religious people. Coming to share more and more of the divine knowledge, we should lose progressively more of our personal identities through fusing with the greater whole 'in order not to miss any of the show', as Nozick puts it; we should 'pass', writes Keith Ward, 'as most theists think, into the wider reality of God,' perhaps eventually coming to 'know God wholly' and becoming 'one reality with God' (1996*b*: 186 and 200; 1996*a*: 239). (Fusing *straight away* with too much of the divine reality would presumably mean an immediate end to personal identity so that no benefits would accrue *to us*. In the Preface to Part Four of the *Ethics* Spinoza remarks that, just as a horse would lose its identity if transformed into an insect, so also would it lose it if transformed into something much more perfect, a human; and this looks fairly obviously correct when the transformations are imagined as abrupt. Describing 'dissolving into Brahman' as following immediately and totally upon leaving this world of ours, the Chandogya Upanishad is right in its conclusion that *we could never know* we had merged with Brahman. There would no longer be any *us* to possess the knowledge. Note, though, that the Taittariya Upanishad instead suggests that the self merging with Brahman retains its individuality although 'changing its form at will' as it roams throughout the divine reality.[7] In his *Mind and Matter* (1958) and *My View of the*

---

7 On the contrast between the two Upanishads, see pages 86 and 87 of Ward 1996*a*.

World (1964), Erwin Schrödinger perhaps straddled the two teachings. In our afterlives we should all of us come to see what physics had, he thought, already told us: that in truth all things are in Hinduism's cosmic mind. There are no separate persons, he wrote—but oddly added that *we* were destined to know it.[8])

Coming to share more and more of the divine knowledge wouldn't have to be like reading ever more pages of a gigantic encyclopaedia or absorbing neatly packaged information from a television set. The parts of the knowledge in which we might hope to share could include knowledge of joys rather nearer to those of skiing, rock climbing, running after butterflies, actually playing chess instead of memorizing every page of *Nunn's Chess Openings,* scientific discovery and invention, composing music, painting, chatting, loving and being loved, and so forth. It is difficult to see why any afterlives inside a pantheistic whole would exclude all new knowledge of experiences anything like these—although it might be knowledge taking forms very hard to picture since it would presumably be possessed without possession of blood, bones, tongues, and eyes. But despite all of us lacking these things, we might hope to recognize dead friends with whom we were reunited, through becoming aware of their ways of thinking.

In addition, there might be experiences of interacting with a divine personality somewhat as pictured by religious thinkers. Such a personality could be associated with the kind of *divine overview of everything* that Spinoza imagined. Interacting with it could be awesome without being terrifying.

If anyone could look forward to any of this, could absolutely everybody? Perhaps viciously unpleasant people wouldn't survive. They

---

[8] Schrödinger agrees with Spinoza that the cosmos is a single existent, yet (Schrödinger 1964: 21 and 22) he views Spinoza as wrong in thinking that you, for example, are simply 'a part, a piece, an aspect or modification of it', truly distinguishable from other people since they are genuinely different parts/pieces/aspects/modifications. Instead, 'you—and all other conscious beings as such—are all in all'; 'this life of yours is not merely a piece of the entire existence, but is in a certain sense the *whole*'. It is just that 'the whole is not so constituted that it can be surveyed in a single glance', which explains why you cannot see what the other conscious beings can. This is, he believes, the Vedantic vision, the message of the Upanishads. Maybe it is, yet it would seem to be self-contradictory.

could fail to be worth preserving whether or not their viciousness was really their fault. But possibly they as well would continue onwards, losing their nastier features with merciful swiftness. Ward suggests that God will give to 'all finite persons' forms which make them able 'to know and appreciate the whole history of the universe' (1996b: 199)—meaning, no doubt, that Genghis Khan and Stalin would come to regret their contributions to it.

Would lions survive? Would frogs or still more primitive animals? If we are unwilling to try answering this, why not at least say that any animals with lives sometimes worth living could well have afterlives also?

The Best and Infinity   4

Pantheists can think of reality as 'of the best possible kind'. This must be understood with great care, though. It is not being suggested that our own universe is the best possible, for the divine mind presumably contains vastly many universes, including ones obeying very different laws. We could hardly expect ours to be the best of them. Again, we are not being asked to praise whatever happens, viewing efforts as pointless. It may be good that the divine mind includes regions such as our universe, regions conforming to laws of physics (although it may also contain much that neither obeys such laws nor is collected into universes). It may be good that no miracles stop us exercising our freedom: the sort of freedom we clearly do possess, regardless of whether laws of physics control our brains. It may be good that humans and animals are not guaranteed by other miracles against fires and plagues. But reality is definitely not perfect in such a fashion that all ethical needs are fulfilled. Fires, plagues, and more exotic disasters (perhaps including destruction of entire galaxies through experiments at very high energies, a possibility discussed in the physics journals), as well as things like murders, could result from very unfortunate conflicts between goods. The good of freedom, for example, and the other goods which, when freedom is used badly, are brought into conflict with the good of freedom. Or the good of causal orderliness, of not living in a world like a drug addict's dream, and the good of avoiding destruction as you ascend the erupting volcano. It is absurd to fancy that such conflicts between goods would never result from our choices.

The difficulties of believing in the goodness of the cosmos are particularly severe when you think that many situations have negative intrinsic value, meaning that each would be actually worse than nothing if it existed all alone. The Privation Theory of Evil denies this, yet it could be judged too counterintuitive. It might be better to theorize (as in Chapter 2) that inside any divine mind situations would always be united with respect to their very existence. Again, a divine mind might be better for not having knowledge with ragged gaps corresponding to such matters as how it feels to be in pain.

Granted that reality as a whole was guaranteed to be infinitely good, we could still have moral cause to improve our segment of it. Imagine infinitely many islands each inhabited by moderately happy people. It would be good to make the people of your own island extremely happy.

Next, picture a world unified in its existence and containing infinitely many islands of happy people. This world could itself be much inferior to a reality made up of infinitely many such worlds. Seeing this, we can well deny that there is just a single unified realm of consciousness or divine mind. We can have grounds for believing in such realms or minds in infinite number. As Chapter 5 will discuss, the cosmos may exist because of an ethical need or requirement, a suggestion made by Plato and others. Now, a scheme of things existing for ethical reasons can have no arbitrary limits set to its goodness.

How to Avoid Idiocy when Saying 'Best Possible'

Suppose that nothing exists apart from divine thinking. Reality must then presumably be 'as good as possible' in a straightforward enough sense. Divine thinking presumably cannot be improved upon. Since, however, it would be idiotic to argue that every event should be greeted with enthusiasm, this matter has to be understood very carefully.

For a start, it isn't being claimed that *our universe* is the best possible. Covering everything worth knowing, the divine thoughts would presumably extend to the structures of vastly many possible universes. The structures would all of them be known in all their details so that (see Chapter 1) those universes would one and all be more than merely possible. Like our universe, they would actually exist inside the divine mind. It would be very odd if ours were the best of all.

If the word 'universe' had to mean Everything, then there could only be a single universe, but modern cosmologists typically don't use the word like that. While they may speak of 'the universe' this is like talking of 'the galaxy', meaning *ours*. In the multiple-universe cosmologies that are nowadays popular, a universe is a collection—separate or largely separate from other such collections—of causally connected things. Universes are often imagined as existing in infinite number and tremendous variety. There is nothing to trouble pantheists here. Spinoza, admittedly, may think the divine mind contemplates only a single universe, ruled throughout by the same basic principles, but this would seem a case of making that mind much less good than it could be. Why not instead contemplate infinitely many universes, and why shouldn't these differ not only in their initial conditions (number and arrangement of particles, etc.) but in their most fundamental laws? Similarly in the case of Spinoza's apparent assumption that the divine mind never considers anything but universe-constituents. Why no contemplation of all possible games of chess, and of every possible move in every possible board game finer than chess, and of all possible fine symphonies, and of all the beautiful theorems that can be proved mathematically, and so forth? Presumably any pantheistic reality that lacked all of this would be less than ideally good. Yet it could be odd to imagine that every possible fine symphony or board game was actually played, every beautiful theorem proved, in some universe or other.

All that a modern pantheist need claim is that our universe contributes at least a little to the goodness of the divine reality. Its contribution might be very inferior when compared with the contributions made by other universes or by regions (scarcely deserving to be called universes) whose constituents weren't ordered and held together by what we would describe as physical laws or causal interactions. Let us assume that talk of how the ingredients of our universe obey physical laws or interact causally is talk of their being arranged in ways such as physicists could hope to capture in reasonably simple equations.

Perhaps a great deal of the divine thinking is devoted to what we might call *hallucinatory splendours* which no such equations could describe. Would those splendours be more splendid than the splendours of universes obeying physical laws? As a pantheist, I need have no views on the subject. I need only suppose that universes such as ours do at least add something worthwhile to the divine mind inside which they exist.

True, many items in our universe are far from splendid. Is a divine mind to have the good of contemplating a system of events which all of them conform to physical laws, thereby coming to have the kind of order that intrigues scientists? If so, then the mind in question cannot have simultaneously, in the very same region of its thoughts, the good of contemplating a system made beautiful with the help of constant miracles. Mountains, woods, and coral reefs may be strikingly beautiful, but the good of having such things in a world could in many places be *overruled* by the good of that world's being law-controlled, so that it contained dreary swamps and lifeless plains as well. Again, couldn't it be better to inhabit a world of danger, as we do, instead of being disembodied brains fed a diet of dreamlike bliss by some gigantic computer? Perhaps so; but this isn't to say that every threat to our happiness is to be welcomed and every disaster admired. Even inside a pantheistic reality all the suffering caused by fire or storm or avalanche could be highly unfortunate—an ugly result of the fact that to have all goods simultaneously, in every single universe which made any kind of worthwhile contribution to that reality, would be an impossibility.

If our universe is part of a divine mind, even *scalar field disasters* (surely the ultimate in disastrousness) could none the less occur at various places and times. In many other universes, no doubt, the laws of physics would make these disasters impossible. In some, perhaps, the disasters would be possible in theory but conditions would be such that they never actually happened. The other universes might then be much better than ours, worthier objects of the divine mind's contemplation; but in addition to contemplating them the divine mind could benefit from contemplating our universe also. Why might scalar field disasters occur? Our universe is believed to have undergone an intensely hot Big Bang. It is thought by many physicists that all elementary particles were originally massless (which photons are to this day). As the universe cooled, a scalar field or fields appeared. Although lacking directionality such as renders a magnetic field detectable with a compass needle, and despite having the same intensity out to distances as great as our telescopes can probe, any such field could make its presence known by giving differing masses to the various kinds of particle with which it interacted. Now, the stability of the field might be rather like that of a ball trapped in a hollow, unable to roll downwards until given a vigorous shove. A possibility taken seriously in the physics journals is that humans might provide such a shove by some experiment at extremely high energies (see Ellis *et al.* 1990; articles discussed at Leslie 1996*a*: 108–22; Rees 1997: 205–7). Collisions between cosmic rays have already released enormous energies inside very tiny volumes, and unless humans managed to produce events still more violent inside equally tiny volumes they would presumably be safe. If they did manage to do it, though, they might 'knock the ball out of its hollow'. An initially minuscule bubble of new-strength scalar field would then expand at virtually the speed of light, changing the properties of particles as soon as it reached them and destroying first the solar system, next the galaxy, then all the nearby galaxies, etc. Well, when a pantheist declares that all reality is the reality of divine thinking, and so cannot be improved upon, this *is not* a case of suggesting that such all-destroying bubbles are impossible.

The very existence of scalar fields, fully stable or otherwise, has yet to be proved firmly. The fact that ours is a world of serious dangers, the human race running quite a large risk of becoming extinct in the next few centuries, can seem obvious on much more prosaic grounds: think of nuclear bombs, germ warfare, the pollution crisis, and so on. A 'doomsday argument' originated by the cosmologist Brandon Carter runs as follows. We humans should be reluctant to accept that our planet was the very first on which an intelligent species evolved in a universe destined to include many billion such planets. Similarly, you and I should hesitate to believe that we existed in, say, the earliest billionth of a human race destined to colonize its entire galaxy. This consideration ought to magnify any fears we have for the future of humankind, moving us in the direction of thinking that our species will quite probably be extinct fairly soon—which in view of the current population explosion would mean that of all humans who will ever have existed up to about 10 per cent were alive today. Despite the very strong possibility that our world is indeterministic, in which case, it could be protested, the number of humans who will ever have existed *wouldn't yet have been fixed* in a way to which we could justifiably appeal, this argument of Carter's acts powerfully against confidence in a long future for human beings. To believe, say, that it is 60 per cent probable that our species will colonize its entire galaxy, and that only one in a billion humans will have lived when you and I did, could seem bizarre.[1] Well, pantheists need not believe it. Pantheism is not a doctrine describing our world as a cosy place.

What is more, it isn't a doctrine telling us we can do nothing to make reality better. That the divine mind cannot be improved upon, and that our universe is just part of it, says that nothing outside that mind could be imagined as able to do anything to improve it. We, however, are supposedly *not* outside it. We are inside it, and our good or bad actions can make our region of it better or worse than it would otherwise be.

In *Micromégas* Voltaire introduces a 'philosophe malebranchiste' who declares, 'It is God who does everything for me, without my interfering'. This might or might not be fair to Malebranche, but at any rate pantheists needn't accept it. Yes, Spinozistic pantheism maintains that all human actions, like all rocks, trees, and stars, are only elements in a divine mind; but it wouldn't follow from this that there were no genuine rocks, trees, or stars, or that humans never genuinely did anything. Likewise, from the fact that the divine mind would in some sense be *eternally the same* it wouldn't follow that everything useful that could be done had, alas, been done already. As noted in Chapter 3, any sense in which pantheism's divine mind (and therefore also all its parts such as you and me) existed unchangingly would be fully compatible with there being a sense in which changes existed

---

[1] For more on Carter's argument see e.g. Leslie 1992c, or 1996a for book-length treatment both of its reasoning (which is controversial, the book discussing a great many attempts to refute it) and of the threats confronting the human race.

inside that mind, human decisions made at any one moment having consequences at later moments. Pantheism doesn't tell us that we can only wait and see what the future will bring instead of struggling to produce good events.

Worthwhile Effort in a Pantheistic Scheme of Things

Those last points are reminiscent of others which keep cropping up in discussions of freedom of the will. A standard position is the 'compatibilist' one that there would be plenty of room for moral effort even in a fully deterministic world: a world which—like a clock—would go through the same sequences as before if it could be returned to one of its earlier states. The sheer fact that you and I were parts of a system entirely obedient to deterministic physical laws would by no means imply that our struggles were pointless. A system developing deterministically couldn't be made any better or worse than it was deterministically sure to be; but so what? *Inside it* decisions would affect just how fortunately or unfortunately the system developed. Deterministic chess computers don't just wait and see whether they will defeat their opponents. Instead they evaluate various possible moves, selecting ones which seem profitable. Suppose the entire world were fully deterministic. Such processes of evaluation and selection would still certainly have effects. When the movements of one billiard ball influence those of another, it isn't because billiard balls *act indeterministically*.

All this would continue to be true, the pantheist can point out, in the case of events that were parts of a system of divine thinking. The events could certainly have effects on other parts of the system. Some of them could be human decisions which could have fortunate or unfortunate results, so that they ought to be made in one way rather than another.

It is the room for worthwhile effort that is crucial here. If any folk disliked classifying efforts made inside a deterministic system as 'moral efforts' but were still willing to see them as worthwhile, then I should be half inclined to encourage them to use language as they pleased. Also, I am not in the business of disproving what is sometimes termed *libertarian freewill*. My own view is that one needs determinism or else some fairly close approach to determinism in order to have a mind that can reason in orderly ways with some hope of controlling its environment usefully. This is the sort of mind that I myself term *free*, at least in the absence of such factors as brainwashing and guns pointed at the head. The world could well have a measure of indeterminism but prima facie this wouldn't help matters. The more indeterminism there is, the less control one has over one's mental workings and over events beyond them, I suggest. Still, some believe in what they call 'absolute freedom'. It involves an absence of obedience to physical laws *and also* an absence of governance by mere chance or by some mixture of physical laws and mere chance. I suspect this is rather like saying 'prime number not below 954 *and also* not above 966'. (There is no such prime number.) I further suspect that absolute freedom in that sense, if possible, would have no particular value. Moreover, if Albert Einstein was right about our world's four-dimensional existence, then this freedom couldn't operate as imagined by most people who believe in it, for they typically insist that the future must be 'absolutely open' when a free choice is made. If it is already true that you are about to have eggs for breakfast, 'having eggs for breakfast just a few minutes further along the fourth dimension' being a description that applies to you rather as 'getting very muddy just a few yards further on' applies to some road, then it cannot be true that your present process of deciding whether to have eggs is free in quite the fashion which these people wish. Suppose, though, that absolute freedom in the sense in question truly is possible and that it would have considerable value. I cannot see why individuals who were elements inside a pantheistic scheme of things would necessarily lack it. I can imagine some pantheist claiming that the divine mind undergoes changes of the absolute kind that Einstein denied, as a prerequisite of having parts splendidly endowed with such freedom, and that the divine thinking would be strictly speaking 'as good as possible' only if all free choices happened to be made in ideal ways. I want to fight only for the following point: that even if all our actions obeyed fully deterministic laws of physics there would still be room for us to make efforts that were worthwhile.

Could this really be the case, though, if all events were parts of a strictly-as-good-as-possible system of divine thinking? Well, why ever shouldn't it be the case? *A system which is the best possible* such as the world imagined in Leibniz's *Theodicy* can be very different from *a perfect system* as imagined by J. M. E. McTaggart and various other British Hegelians, a system in which the need for one good never genuinely conflicts with the need for another. (McTaggart wrote that, 'fortunately, the attainment of the good does not ultimately depend upon action', for otherwise it could be 'rather alarming to think that there were cases in which we did not know how to act'.[2]) The ability to make up one's own mind about what to do—which in a fully deterministic system would be somewhat like a chess computer's ability to make up *its* mind—might be a great good; the good of not murdering people might be another; and these two goods could be in potential conflict. What one then ought to do (yes, even in a fully deterministic world) would be to bring the two goods into line with each other by deciding not to become a murderer.

Someone could raise the following objection: 'Might the world not have been guaranteed to be such that right choices were always made? In the case of a fully deterministic world, guaranteeing right choices would be a matter of creating the world's particles in suitable arrangements. The unfolding pattern of events would then never include bad choices. In an indeterministic world, on the other hand, a world in which physical laws failed to dictate every detail of every event, all that would be needed would be this: that the events in people's brains were guaranteed to occur in such ways that bad choices never occurred. And why shouldn't this be guaranteed? After all, if good choices were physically possible then all that would be necessary, at least in any pantheistic scheme of things, would be for the divine mind to think of those choices as the actually occurring ones!' However, the pantheist who asks people to choose good things rather than bad ones can rebut any such objection. For (1) it is unclear that there could be any initial

---

[2] This comes from McTaggart 1901: s. 135. The desperate attempts of British Hegelians to find grounds for recommending moral efforts inside a strictly perfect universe are discussed at Leslie 1979: 97–9.

arrangement of a deterministic world's particles which would ensure, say, that every single one of the billions of coins tossed in that world would land heads, let alone that every single brain in it would operate virtuously. And (2) if the world were instead indeterministic, it might be doubted whether decisions could count as *choices* in any interesting sense if they were guaranteed always to be virtuous. (3) Assuming, though, that in *some* worlds, whether deterministic or indeterministic, everybody always chose virtuously, the divine reality would still include many other worlds as well: worlds such as ours.

Although there are leading quantum theorists who continue to deny this, it is nowadays usually accepted that quantum physics makes our universe *in*deterministic. Its indeterminisms are thought of as being evident only at submicroscopic levels. In everyday life we can often be fairly confident about what is going to happen, rather as you can be confident on tossing thirty tons of coins that about fifteen tons will land heads. Still, some very queer events are possible, such as suddenly finding you have quantum-tunnelled through a brick wall. Now, pantheists might believe that the divine reality included all possible indeterministic universes that started off precisely as ours did. Some very rare universes could then be ones in which coins never fell heads though many billion were tossed. Some might be ones in which brains always selected the best possible ways of behaving. Some might actually be universes in which, despite the presence of trillions of potential victims, the rule of physical laws never led to any such disasters as death by fire or plague or falling downstairs. It might be that in a few nobody suffered the slightest injury such as getting pricked by a thorn. Yet against this background it would still always make sense to try to choose well rather than badly. Suppose, instead, that *just the one* indeterministic universe existed. Trying to choose well could favourably affect the chances of getting beneficial results, obviously. Now, when there existed all possible indeterministic universes that had started off precisely as ours did,[3] then trying to choose well could equally

---

[3] Many-worlds quantum theory, sketched by Hugh Everett in 1957 and nowadays popular among quantum cosmologists, maintains that every possible world that obeys the laws of quantum physics and that started off as our world did *actually exists*. Leslie (1989a: 84–91) gave various reasons for taking this seriously. I have since come to see that

favourably affect the statistics of the ensemble of universes. It could favourably affect the probability that the particular universe in which one found oneself would be a universe in which beneficial results were found as well.

In short: whether or not our world is deterministic and whether or not it is only one from a vast ensemble of very similar worlds, there is no getting away from the fact that efforts to produce good results can be worthwhile. Pantheists and non-pantheists alike can accept this truth.

'Yet', it might be protested, 'wouldn't your pantheism's divine mind be sure to contain infinitely much good regardless of how we behaved? After all, we are supposed to be infinitesimally tiny parts of the mind in question. Our entire universe, we are told, is probably just one of infinitely many that exist inside it.'

To this I can only reply that the infinite good of God's mind would by no means make it quite all right to murder Mr Brown, on the excuse that reality as a whole would continue to have infinite goodness after he had been murdered. This point would be totally unaffected by whether Mr Brown was himself just part of God's infinite thinking. Although eager to classify myself as a 'utilitarian' intent always on maximizing goodness, this doesn't at all mean I ought to be satisfied by any good that could be called 'infinite numerically', the sort of good that would be present in an infinite set of lives each having a value barely above zero. When quantity of goodness was guaranteed to be infinite, *quality* could still be important. Suppose there were infinitely many islands each carrying people who led fairly contented, fairly

a common way of presenting the theory runs into a fatal difficulty. In the way in question, our world (and every person in it) is pictured as *branching* into more and more 'versions' as time goes by. One aspect of the difficulty (discussed in Leslie 1996d) is that the vast majority of the versions into which you yourself branched would then exist at moments when they were just about to die. This could give you overwhelming grounds for believing your own death to be imminent, regardless of how safe and healthy you appeared to be. (If you accept the theory as presented but are not convinced by this reasoning, so that you remain confident that you will be alive at the end of the month, then why trust your calendar's assertion that it is only Monday 11th today? The spatiotemporal map of the world which you think correct shows vastly many more versions of you observing Tuesday 12th. Probability theory then suggests that you should consider the calendar's evidence as outweighed. Your protest that the versions of Tuesday 'aren't there yet' is invalid, I would argue, whether or not Einstein is right about the world's four-dimensional existence.)

interesting lives. If waving a magic wand could give all these people extremely interesting, extremely happy lives, then the wand ought to be waved. And improving the lives of people on just a single island could itself be worthwhile: *exactly as worthwhile* whether or not any other people existed anywhere. That they existed in infinite number couldn't change this.

Spinoza made the world a better place. At times he acted courageously in support of others. It is a shame, then, that his official doctrine might appear to be that we should admire all of the world's events equally, reflecting that each is a necessary element in the divine reality. (For instance, the Note to Proposition Forty-Nine of Part Two of the *Ethics* encourages us to 'wait for and bear with equal mind all forms of fortune, because all things follow from the eternal decree of God'.) But luckily he is far from consistent about this. His writings are full of moral advice—of passages suggesting that some of the world's events are much finer than others and that instead of waiting to see what will happen we ought to try to make it happen admirably.

Pantheism and the Privation Theory of Evil

The theological Problem of Evil—the puzzle of how to square belief in divine goodness with recognition of the world's disasters—looms largest when the intrinsic value of various things is judged to be *strictly negative*. Now, the ordinarily accepted view is that many conscious states do have intrinsic value of that unfortunate type. When an Augustine, an Aquinas, a Leibniz, or a Spinoza defends what is known as the Privation Theory of Evil, this comes as a real shock to many people. Expressed with analytical precision, the theory can seem to run as follows: that no things whatsoever, not even states of extreme pain, are *in themselves actually worse than nothing,* that is, such that they would be bad if they were the only things in existence. But can anyone genuinely have believed this? Isn't it self-evidently ridiculous?

I have very little confidence in my own intuitions here. Like almost everybody else's, my initial reaction is that severe headaches are all the proof we need of the Privation Theory's wrongness. Still, may such a reaction not be just a product of Darwinian pressures? Why trust it as a guide to the truth? After all, privation theorists can agree with everybody else about what's better than what. Where they differ from most of us is in their placement of the zero on the scale of values. To them, not even the worst situation has a value below zero when considered just in itself. Well, can this be self-evidently wrong when there are some exceptionally intelligent philosophers who think that the whole idea of anything's having intrinsic value—not just of a negative sort, but of *any* sort—is only some strange delusion? The Privation Theory comes as much less of a shock to me when applied to physical rather than to mental pain, perhaps at the death of a loved one. However, wouldn't some say instead that it was physical pain that was obviously intrinsically evil whereas sorrow at a death was an intrinsically fine appreciation of the beauty of the dead person's life? And how could anyone know for sure who was right here? Conceivably the Privation Theory is correct no matter to what kinds of pain it is applied.

In a letter of 1665 in which (as is traditional among privation theorists) he compares all evils to blindness, commenting that this is nothing but the absence of the good of sight, Spinoza might certainly be considered grossly mistaken. He could seem to lump evil situations together with holes in buckets. 'The reality is the buckets', we might want to say, 'while the holes are just local absences of bucket'; yet when we think of such things as acute miseries, shouldn't it be obvious that these are more than mere absences of joy? Still, what if Spinoza's main point were that a miserable state of mind could never be *such that its existence all alone would be worse than utter emptiness, the presence of a blank*? Would this point be preposterous enough to ruin his entire philosophy?

Possibly it would be, were it in fact essential to his pantheism. Notice, though, that the idea of intrinsic value that I have been using here, an idea which came to be described rigorously perhaps only in the twentieth century (by G. E. Moore), meshes rather poorly with Spinoza's 'monistic' way of thinking. In Spinoza's eyes, the world's parts are never *such as could exist all alone* so that it could be asked in all seriousness whether each was better or worse than a blank. Spinoza's divine mind is unified in its existence.

All the same, might not any Spinozistic pantheism remain in grave difficulties? Even if no parts of the world are such as could exist in isolation, might not this feat be performed by various things really very much like them? The colour of a brick, its length, and its hardness, are all abstractions very obviously incapable of existing in isolation. A miserable conscious state, or at any rate the entire individual who is in that state, is much less obviously so. And couldn't it be essential in ethics to evaluate this or that as if it could exist in isolation, distinguishing the value of some situation *in itself* from the 'instrumental value' it possessed through producing various effects? How else could we discuss, say, the morality of surgical operations which cause pain but save lives? Granted that everything existed inside a divine mind that was unified in its existence, evaluating things *each on its own* would be somewhat artificial, but not too much so. 'Why on earth,' we could ask, 'would the divine intellect devote any of its powers to contemplating precisely how it feels to be miserable?'

Perhaps the best strategy for pantheists like me is as follows. By all means let us throw doubt on the idea that there exist things each of which would be worse than a blank, were it to exist all alone. Let us insist both on the fallibility of our intuitions about things like headaches and on the point that, inside the divine mind in which we believe, individual things are never such as could exist each on its own. But let us also ask whether anyone can be confident that such a mind, contemplating a possible world obeying the physical laws that our world obeys, *would have been better for being ignorant of many of its details*, thereby ensuring that those details remained confined to the realm of possibility instead of taking on actual existence. Would the divine knowledge of any such world be finer if it were ragged or incomplete like a jigsaw puzzle with various pieces missing? For example, if it were confined to such affairs as Mozart's experiences of composing great music, with large gaps with respect to such others as what-it's-like-to-be-Hitler-enraged-by-losing-his-war? If God had to be defined as a person who had chosen to have truly complete knowledge of all our world's events, could we then firmly disprove God's existence by arguing that he would have chosen to know exactly how it felt to be a human in pain, which would be absurdly masochistic? Surely the answers to such questions are far from obvious.

Infinitely Many Divine Minds

'The divine mind' might be words like 'the galaxy' or 'the planet'. What could be meant would be *our* divine mind: the one of which we were supposedly parts. There would be no implication that there existed only a single mind worth calling divine. To be divine, a mind would presumably have to be infinite; but remember, 'infinite' need not mean 'including everything'. It is traditional to call God infinite whether or not you accept the pantheistic position that your mind exists inside God. A mind which knew infinitely much could be an infinite mind despite there being many other minds outside it, and it could be infinite even if those other minds were infinite as well. Bear in mind, too, that any creative factor might be expected to operate repeatedly, which could be particularly obvious when the factor in question was the Good of Plato. If the reason why a divine mind exists is that it is eternally ethically required that there be such a mind and that ethical requirements, when not overruled by other ethical requirements, are themselves creatively effective, then wouldn't one expect the existence of infinitely many such minds *because this was best*?

The Platonic theory that ethical requirements can sometimes themselves ensure their own fulfilment is the topic of the next chapter. But first, would it really be better for there to exist more than one divine mind? And would it be better for there to be an infinity of such minds, rather than some finite number?

Some philosophers consider that, long before reaching the infinite, there would be as much good as there could ever be. How many humans, for example, would it be good for our universe to contain in its spatiotemporal entirety, or how many intelligent beings of any sort? A few billion would be quite enough, they think. Anything more would be boring repetition. But this seems to me a very strange way of thinking. (1) It appears to forget that each new human or other intelligent being wouldn't find life less exciting just because many others had lived their lives previously. How could falling in love be less wonderful today because the Romans had fallen in love earlier? (2) Again, suppose that the Privation Theory of Evil were wrong and that a miserable life had a

148

value that was at least slightly negative. Without being insane a man might persuade himself that many trillion happy lives, or even infinitely many, would be really no better than only a few billion; but who but a raving madman would judge that many trillion lives of negative value, or infinitely many of them, would be *no worse* than a few billion? Yet if the *negative* value of a group of miserable lives became ever greater as that group increased in size, whereas the *positive* value of a group of happy lives would in contrast reach some upper limit, then it would follow that a single island bearing one miserable person and ninety thousand happy ones could be something good while some gigantic number of islands precisely like it would instead be a disaster, and an infinite number of them an infinite disaster. Now, how could that make sense?

The moral would seem to be that the goodness of a scheme of things cannot reach any upper limit.[4] The good, no matter how immense, of any one existent or group of existents would never make it any less ethically necessary for there to be others as well. If it were good for there to be one mind that knew infinitely much, it would then be still better for there to be two such minds, and better yet for there to be three of them; and so on; just as if there were one mind of infinite negative value, then it would be still worse if there were two, etc. (If there were two minds of infinite goodness, destroying one of them would be a great evil. Would this be because once any such mind had existed for however brief a period, it would then be a great good that it should continue in existence although *its coming into existence* wouldn't have been good at all? Well, why should we believe anything so strange? Consider once again the case of two minds of great or infinite negative value. Who would dream of arguing that annihilating one of them would be good yet that its coming to exist could not have been bad?)

It is traditional to assume that there would be only a single divine mind. Western theology, anyhow, has no multiple divinities. All I want to argue for is the idea that there are infinitely many minds each knowing infinitely much that is worth knowing. I call all of them divine, but if other people preferred to use the words 'divine

---

[4] For further arguments favouring this conclusion, see Parfit 1984: part IV.

mind' only of something believed to be unique, then why go to war with them? My language would simply differ from theirs. Still, how about the notion that multiplying infinitely knowledgeable minds would make each of them less glorious and therefore less deserving of being called 'divine'? This seems well worth fighting against. How could a mind's knowledge be any less intrinsically worth having through being had by some other mind as well? How could existing all alone contribute to divine glory? Perhaps there couldn't be two omnipotent devils because each would want to limit the power of the other, but divine minds surely wouldn't be in that unpleasant boat.

Would Divine Minds Need to Differ?

If there were many minds each knowing immensely much, minds of the type I prefer to call divine, would they necessarily know different things? Or would it be good, at any rate, if the things they knew were different?

Let us take the second question first. Given that an angel is better than a stone, must two angels be better than an angel plus a stone? 'Not necessarily', ran the medieval answer. 'An angel plus a stone would have the good of variety.' Now, what if a Platonic creation story were correct, vastly many immensely knowledgeable minds existing because this was the best possible situation? Granted that a first mind, a mind knowing everything worth knowing, was as fine as any mind could be, might a mind that knew marginally less be most ethically desirable as a second mind, through the need for variety? Might a mind of still greater ignorance be the best available third mind, and so forth, so that one perhaps got a Great Chain of Being stretching all the way down to some mind that knew hardly anything? These strike me as absurd suggestions. Certainly variety can be a good. A divine mind's knowledge could owe much of its value to its being *immensely varied* knowledge. But this doesn't say that a second mind would be a better addition to reality if its knowledge differed from that of the first. Variety would not be desirable here.

Variety might be necessary, though. The Principle of Identity of Indiscernibles might necessitate it. This is the principle that supposedly prevents there being more than one angel at the very tip of an infinitely sharp pin, granted that angels (for the purposes of this thought experiment) are point-particles, beings who have no size, and are identical in all properties apart from spatial position. Not even omnipotence could then create two angels at one and the same point, says the principle. Existing things cannot differ 'in number alone' or 'solely in their being'. Indiscernibles—things identical in all properties—must be numerically identical, one and the same, like Napoleon and Bonaparte. 'Isn't this utterly obvious?', it might be asked; 'for how on earth could it be true that a thing satisfied such and such a complete description, and also true that a second thing satisfied a precisely similar description? If the descriptions were no different, how could they possibly differ in the things they picked out? What would you think of a drunkard who insisted there were two identical lampposts, entirely overlapping?'

I answer that such rhetorical questions establish nothing. Those who reject Identity of Indiscernibles can ask in return, 'Why on earth would things numerically different—things different in their being, that is, in the fact that the existence of the one wasn't one and the same reality as the existence of the other—necessarily be things such that different descriptions could pick them out? Why should *existing separately* demand difference *in description*? Since when has all real existence depended on describability, let alone on being describable in a fashion that picks out uniquely?' And while the theory about the lampposts would be silly, photons seemingly *can* occupy one and the same volume. Instead of colliding and bouncing back, two photons can pass right through each other like waves on a lake. Besides, in the case of infinitely many separate minds, each knowing everything worth knowing, there would be no question of occupying one and the same volume. The minds presumably wouldn't even exist in any spatial relationship to one another. (In some cosmological theories, universes occurring as quantum fluctuations are said to exist each in a space of its own. Plausible or not, this could be thought no offence against logic. And why should logic oppose the idea that two such quantum-fluctuational universes chanced to be precisely alike? If a universe of a precisely specified type quantum-fluctuated into existence, could it possibly be a law *not of physics but of logic* that this was unrepeatable?)

The Best and Infinity

Suppose, however, that multiple divine minds did have to exist side by side in space of the kind familiar to us. Let us try to imagine them as absolutely identical in all their intrinsic qualities, and as in some totally symmetrical arrangement. Let us try saying that each possesses its consciousness of its own existence, its own thoughts, *and not* the consciousness or the thoughts of the others. Well, people who support Identity of Indiscernibles will not let this pass unchallenged. They deny that things could differ 'in number alone', that is, merely in their existence and not in their qualities. (They aren't saying just that two minds that were precisely similar would still be 'non-identical' in the trivial sense that each mind wouldn't be the other one.) They must accordingly reject the situation described. If either the symmetry of arrangement or the similarity of intrinsic qualities were defective to some infinitesimal degree, then they would find the situation acceptable, but their theory is that perfect symmetry combined with identical intrinsic qualities is an impossibility, a contradiction. Well, does that sound sensible?

Next, try to imagine four completely homogeneous, precisely spherical spheres made of precisely similar material. Each is placed at one of the corners of a perfect square and there is nothing else in existence. It could seem there were differences here. Taking any one sphere, wouldn't two of the others be nearer to it than the third was? None the less, Identity of Indiscernibles would maintain that the situation was utterly impossible. Dent one of the spheres infinitesimally, or reduce infinitesimally the symmetry of their placement, and the situation could occur; yet as things stand it would be an offence against logic, says Identity of Indiscernibles, since any description satisfied by any one sphere would be satisfied as well by the three spheres allegedly distinct from it. (The description 'sphere furthest from *this one*' would be banned because nobody would exist to pick out which sphere was *this*. The description 'sphere furthest to the left' would be banned as well, for in a truly symmetrical cosmos there is absolutely no way of making the distinction between left and right.) Now, can that make any sense?

Here is a yet greater paradox for Identity of Indiscernibles to swallow. Try to picture a cosmos consisting just of three qualitatively identical spheres in a straight line, the two outer ones precisely equidistant from the one at the centre. Aren't there plain differences here? The central sphere must be nearer to the outer spheres than these are to each other. Identity of Indiscernibles shudders at the symmetry of the situation, however. It holds that the so-called two outer spheres must really be only a single sphere. And this single sphere, which now has all the same qualities as its sole surviving partner, must really be numerically identical to it. There is actually just one sphere! And next, the two halves of that sphere, being once again the same in their qualities, must likewise be numerically identical so that we have only a hemisphere, which in turn becomes a quarter-sphere, and so on, until all we are left with is an infinitely thin splinter, a line, which must in the end shrink to a point. Does *that* make sense?

Imagine, finally, two universes which are the only things in existence and which differ in one respect alone. At the very centre of the first universe there sits an electron while in the second the corresponding electron is differently placed by a millionth of millimetre. Suppose that these universes are developing in a time that flows in an absolute fashion: not the sort of time in which Einstein believed, but a time in which the situations that people describe as 'existing in the past' have in fact lost their existence absolutely. What if the two universes appeared on the point of becoming identical in their every existing property, through one of the electrons moving slightly? Would one of the universes have to vanish, or would the two of them be replaced by just a single universe without either of them vanishing? This could strike us as altogether too odd.

Perhaps, though, such considerations don't entirely refute Identity of Indiscernibles. What if it is correct after all? There then cannot be infinitely many identical minds each knowing everything worth knowing. Would some people now say that there could be only the one divine mind, a mind knowing absolutely all that was worth knowing, while any other minds would have to be at least marginally inferior and so 'not truly deserving to be called divine'? As Chapter 2 indicated, this wouldn't trouble me greatly. There could still be infinitely many minds of the type I myself would call divine, minds knowing immensely much. Apart from just one of them, each of the minds could be ignorant of a single insignificant fact among the perhaps infinitely many facts worth knowing: a different fact in the case of each mind. A situation of this kind would be impressive enough.

At the same time, let us concede something to Identity of Indiscernibles. Could we argue that any infinite minds that were identical would preferably be fused so that they were present together inside a single existent, 'thus preserving the unity of the divine'? No, because inside any single existent Identity of Indiscernibles would indeed apply. For an analogy, consider a stone which has mass, solidity, a length of 5 inches, *and also* a length of 5 inches. That's nonsense, isn't it? A length of 5 inches can be had by any one stone only once.

A Cantorian Question

If there were infinitely many divine minds, at what Cantorian level of infinity would they be? As discussed in Chapter 1, Cantor is usually taken to have shown that there are endlessly many infinite numbers each greater than the one before; so wouldn't it be nonsense to speak of *a best situation* in which there existed as many divine minds as possible? I think not. Cantor's conclusions may be valuable to mathematicians considering such abstract entities as *points*. (Are there fewer points in a mile than in a cubic mile, or in an infinite space?) They are much less clearly applicable to books, to people trying to fit into hotels whose every room is filled already, or to minds worth calling divine. Besides, I noted that Cantor himself appears to have accepted that his reasoning worked only for things considered as collected into *sets,* technically defined. He thought he *hadn't* proved that, regardless of how infinitely much God knew, some other being could know more 'at a higher level of infinity'. I suggest, then, that the question 'At what level would the infinity of their number be?' either wouldn't apply to an infinite collection of divine minds, because they wouldn't be the sorts of entity for which such Cantorian issues arise, or else ought to receive the answer, 'At the level of Cantor's Absolute Infinite, which he described as *not subject to further increase*'.

Theologians please note that any difficulties of this area are not the fault of the suggestion that divine minds exist in infinite number. They would apply equally to the question of how many turnips an omnipotent creator could create.

Necessary Divine Existence

Suppose we followed the empiricist principle that, so long as it does not positively conflict with experience, a thing should never be looked on as surprising. We might then simply accept a picture of infinitely numerous, infinitely rich realms of consciousness or divine minds, one of them containing the structure of our entire universe. Still, a liking for simplicity could encourage us to reject the picture. Accepting it might also be thought too much like believing in Father Christmas. What we appear to need is some plausible creative principle leading to the existence of infinitely much worthwhile consciousness. The principle might be provided by the Platonic idea that the good is the ethically required, and that ethical requirements, when not conflicting with other, stronger ethical requirements, can themselves be creatively powerful. Believing this could be preferable to various alternatives: maintaining, for instance, that infinite situations are somehow 'really simpler' (even than a blank?) or else that all possibilities, including infinitely knowledgeable minds, must exist somewhere (the 'modal realism' discussed in Chapter 1). Or conjecturing, maybe, that something or other—perhaps a divine person whose acts of will are directly effective or perhaps rather a bad cosmos—exists for no reason whatever.

A Platonic or Neoplatonic story could include elements like these: (1) It would be impossible to get rid of all realities since some are Platonic realities: mathematical facts, for instance, or the truth that the absence of a world of torment would be ethically required whether or not anybody actually existed to think about this, let alone to have a duty to do anything about it. (Various approaches to ethics popular today, such as 'prescriptivism', leave no room for any truth of that kind. In the eyes of prescriptivists, labelling things 'good' amounts merely to calling for actions that favour such things.) (2) Similarly, the presence of various situations could be ethically required in an absolute way, whether or not anybody existed yet. (3) No logician can prove that what is supremely ethically required must actually be the case. All the same, there is no conceptual confusion in the Platonic notion that an ethical requirement, when not overruled by other ethical requirements, could by itself carry responsibility for the actual existence of something. What is more, it might carry this responsibility necessarily, because not all necessities are logically provable. Asking what 'gave' such a requirement its creative power would be like asking what 'made' the experience of red more like the experience of orange than like that of yellow. In each case the right answer would be 'Nothing'. (4) The existence of an immensely knowledgeable mind—a mind perhaps well worth the name 'God' even if there were other such minds as well—could be an eternal consequence of its ethical requiredness. Since the requiredness would have its source in that mind's own tremendously rich nature, it could be misleading to say that this 'rendered something outside God responsible for God's existence'.

Efforts to Explain All Existence

Philosophers priding themselves on their 'empiricism' usually say that only experience can tell us that something needs to be explained. They then often classify the sheer existence of the cosmos as a clear enough case of something *not* needing explanation, for how could we ever have experienced the absence of a cosmos? Or how could we know that the cosmos leapt into being thanks to the influence of some empirically detectable factor which wasn't an existing thing?

If they truly did accept that what doesn't conflict with experience should always be treated as unsurprising, then there would be nothing to astonish empiricists in the existence of infinitely many minds of the kind I call divine. At least, there would be nothing to astonish them if, as I argued, each divine mind would know the structures of countless possible universes obeying physical laws, the existence of our own universe being simply the knowing of its structure by one such mind. A properly developed pantheism is not a rejection of the world as we know it. It pictures our world as actually existing inside a divine mind, and it pictures us as getting to know facts about this world in the ways recognized by scientists. Our experiences wouldn't themselves tell us that we were elements in a divine mind, let alone that such minds existed in infinite number; but equally they wouldn't say that this was wrong.

Nevertheless, might not acceptance of it be sheer wishful thinking, like a belief in Father Christmas? Or wouldn't a divine mind be an offence against simplicity? Well, the pantheist's 'belief in Father Christmas' wouldn't involve expectations different from those of other folk, at any rate with respect to this present life as distinct from any life after death. To suppose that our knowledge of life in a world ruled by laws of physics is simply a divine mind's knowing, in some tiny part of its thought, just what it is like to live in a world of physics, doesn't say that the laws of physics could be expected to break down constantly so that, for instance, virtuous people would never go hungry.

Again, perhaps not even an infinite collection of divine minds would offend against simplicity. In science and in philosophy the simple numbers are zero, one, and infinity, we might well think. In contrast, *exactly five hundred and ninety-six divine minds* could be a fine example of implausible complexity.

Would any divine mind be hugely complex through knowing immensely much? Well, consider how simply its knowledge could be described: *it would know everything,* or else *it would know everything worth knowing.*

Points of this type are exploited by Richard Swinburne. Swinburne writes that God's existence is 'a brute fact', something inexplicable 'not in the sense that we do not know its explanation, but in the sense that it does not have one'. But believing this isn't believing in anything very complex, he claims. God is 'the simplest kind of person that there could be'. God's capacities being 'as great as they logically can be', the divine existence is much simpler than yours, for example. The hypothesis that God is limited in his power, say, or in his knowledge, is more complicated than the hypothesis that God's power and knowledge are unlimited: compare how the hypothesis 'that some particle has zero mass, or infinite velocity, is simpler than the hypothesis that it has a mass of 0.32127 of some unit, or a velocity of 301,000 km/sec' (Swinburne 1970: 92–4). In so far, then, as one can throw light on something by showing it to be simple, we might have here some means of throwing light on God's existence and then perhaps also on that of all other things if they existed (as Swinburne thinks they do) because God willed them to exist. God's existence, Swinburne writes, would be 'not merely *an* ultimate brute fact, but *the* ultimate brute fact'—a brute fact on which 'everything else in the universe' depended, so that the universe formed a whole far simpler than any ragbag of entities existing for all kinds of different reasons.[1]

Swinburne suggests that God, apart from having willed our universe to exist in the first place, fixing its physical laws in the same act of will, keeps it in existence by willing this as well. Nails and hammers, stars and seas and humans, would vanish at once if God did not will their continued existence. But not everything is dictated by physical laws, Swinburne thinks. God disturbs the natural order so as to produce our embodied consciousness. Also, we sometimes have absolute freedom to decide just what will happen. Yet here, too, simplicity can be found, for all of the resulting world is a product of *agent causation*: the causation originating in our absolutely free acts of will, plus the causation originating in the absolutely free divine will.

Is this fully satisfactory, though? While knowing that I choose things and do things, it seems to me that I lack any clear acquaintance with Swinburne's agent causation. Working without such causation and without the need for God to intervene to bring about embodied consciousness, a world might be much simpler. And even granted that

---

[1] Swinburne 1977: 267. See pp. 139–40 for Swinburne's view that God brings about the continued existence of the universe and its obedience to natural laws.

Swinburne's divine person would be in some respects a very simple being, for instance through fitting the highly compact description 'knows everything', there would be other respects in which this person could be considered very complex: infinitely complex in knowledge, for one thing. And anyway, aren't absolutely all *persons* far more complicated than stones or atoms or utter emptiness? Yes, the idea of bringing things about by mere act of will is in a way quite simple, and not at all obviously foolish. (In casinos everywhere, people try it. *That* may be foolish, but why couldn't God do what they cannot?) Still, one might like to have further insight into why the process worked. To be assured that it can work for no reason whatever in God's case, and that things are simplest that way, and that likewise the infinitely knowledgeable divine mind exists for no reason whatever, this also being simplest, can be disturbing.

In addition there could be the following difficulty, at least for those who reject pantheism. If God truly is an omnipotent person, able to create absolutely anything except round squares, torments of great intrinsic worth, puzzles too hard for him to solve, etc., then why does he create a world of strictly limited beings like you and me? Why not instead create an infinite number of persons each having all the features (or if that is impossible because of Identity of Indiscernibles, then *almost all* the features) which make his own existence so immensely good?

Yet is it possible to escape brute facts in this area? David Lewis's attempt to escape by theorizing that all possibilities simply have to exist somewhere, *the actual* being what exists *in our world* while *the merely possible* is what exists *elsewhere,* has struck most philosophers as unsuccessful; and besides, could we trust inductive reasoning if Lewis were right (see Chapter 1)? Another attempt, the theory that an absence of all things would be a contradiction, is also generally considered misguided. So shouldn't we perhaps look for the simplest available brute facts, and then believe in those? Why not just a brute fact *that there exists the universe we know*? It would be what I'd opt for if forced to choose between it and a brute fact of the existence of a divine person able to create everything else by sheer willpower, or a brute fact that there exist infinitely many minds each knowing everything worth knowing.

The difficulty is of getting from the abstract realm of Platonic facts—facts which absolutely have to be the case, like the fact that IF two sets of two apples ever were to exist THEN there would exist four apples or the fact that apples (unlike married bachelors) are logically possible—to the kingdom of existing things. In the Platonic, necessarily real realm, could there be any factor that required the existence of something in more than just an IFy-THENy manner? Might the something be an omniscient, omnipotent person, or maybe a collection of infinitely many minds each knowing everything worth knowing?

The Platonic Theory of Why there is Anything in Existence

Platonic realities may be somewhat shadowy. They would not normally be called 'existing things', like fruit or neutrons or electromagnetic fields or states of consciousness. But this fails to make them unreal. Regardless of whether apples had ever existed, there would be no internal contradiction in the existence of apples; they really would be logical possibilities; and it would actually be the case that in any two sets of two apples there would have to be four apples. These matters, and infinitely many similar ones, do not owe their reality to the actual presence of any people or objects. They are real—necessarily and eternally—just through themselves being what they are. They do not have to be thought about to be real. They did not become real only when languages arrived so that they could be described. If nothing at all had existed, it would still have been the case that apples (in contrast to round squares) could have existed without contradiction, as could a universe like the one we inhabit. The logical possibility of our universe did not depend on the actual existence of this universe or of anything. Now, some Platonic realities are realities of *ethically required existence*. Whether or not there was anything in existence, or ever would be, the existence of various things could be ethically needful. An absence of all existents would be in a way tragic because there might have been a good situation instead. And a situation sufficiently good—one which wouldn't be rather a pity because something far better could be there in its place—could perhaps have an existence *required not just ethically, but with creative effect*. Without contradiction or other absurdity, perhaps, the goodness of something might bear responsibility for that something's being more than merely possible.

This creation story is suggested in book 6 of Plato's *Republic*. The Form of the Good, we are told, 'is itself not existence, but far beyond existence in dignity' for it is 'what bestows existence upon things'. Here are ways in which the story could be expanded:

1. An absence of all existents truly would be tragic. Don't protest that it could be in some respects fortunate, for I am not concerned to deny this here. It, too, could be true. Had there been nothing in existence, an immensely important ethical requirement could still have been fulfilled, namely, the requirement that there *not* exist a situation of immense negative value: perhaps a world consisting solely of people in agony. (The Privation Theory of Evil could be wrong in its claim that things never have negative intrinsic worth.) Yet this just helps show that even in a blank—an absence of all existing things—there could be real ethical requirements. You don't have to have *somebody actually there*, somebody able to shudder at the thought of a world of people in agony, somebody burdened by a duty to prevent its existence, to *give reality to* any ethical need for it not to exist.

2. No examination of concepts can prove that whatever is ethically required must actually be the case, or even that the ethical requiredness of some one entity (a divine mind, perhaps) could guarantee its existence. For a start: as the previous chapter emphasized, ethical requirements could often conflict with one another, guaranteeing that many of them *wouldn't* be fulfilled. Further, not even a supremely good situation could be proved to exist just by our noticing that 'having goodness' means 'being something whose existence is required ethically'. The concepts of *ethically required existence* and of *existence required with causal or creative success* truly are two concepts, not one. So far as mere concepts are concerned, absolutely any ethical requirement might fail to be fulfilled.

3. None the less, there is no conceptual absurdity in the Platonic theme that an ethical need, alias an ethical requirement, might *by itself*

be responsible for a thing's existence. As Keith Ward writes, 'Of course, that something is desirable does not entail its existence'—for 'entail' means *make demonstrable by logicians,* as in 'being a loving wife does entail being a loving woman'. 'But if', he continues, 'there is something which, as Aristotle has suggested, cannot exist otherwise than it does, the best reason for its existence would lie in its supreme goodness.'[2] Plato's theory about why the cosmos exists is a strictly speculative one. It could very easily be wrong. But that does not make it a silly theory.

Think here of what people have often held about divine commands or acts of will. Swinburne is by no means alone in thinking that these, which are requirements that various situations shall exist, would be creatively effective even if those situations were 'marked out for existence' or 'required' not in the least in themselves, but only through the divine desires. I detect no logical absurdity in such an idea. So far as logic was concerned, things might leap into existence whenever a divine being wanted them to do so, it being a totally reasonless fact that it was *with creative success* that the being willed their existence. Now, all the less do I detect any logical absurdity in the idea that an ethical requiredness possessed by some possible thing or things—possessed necessarily, as a matter of Platonic fact—was by itself creatively sufficient.

4. What would give such requiredness its creative power? The Platonic theory is that nothing would give it power in the sense of standing outside it and *making it* creatively effective. It would be the ethical requiredness itself that was creatively effective. (Compare the case of asking what *gives* to a red afterimage its ability to be nearer in colour to an orange one than to a yellow one. The right answer would seem to be that nothing at all gives it this ability, just as nothing at all gives to two green afterimages their ability to be nearer in colour to each other than to a purple one; they just are nearer in colour, and

---

[2] Ward 1996a: 196, with reference to the Platonic creation story as told in my *Universes* (1989: ch. 8). For Aristotle's views, see his *Metaphysics* in particular. 'The cause of all goods is the Good itself' (*Metaphysics* 985a). The Good is 'the First Mover' (1059a), otherwise known as God. God 'exists necessarily, and inasmuch as he exists by necessity his mode of being is good' (1072b); his mental life is of eternal blissful contemplation.

that's that.) The fact that the creative power *wasn't logically demonstrable* would not prove that it couldn't be necessary in an absolute manner. The doctrine that the only absolute necessities are logically demonstrable necessities, like the necessity of there not being round squares, could be considered a mistake. It might, for example, be necessary in an absolute manner that a world of a certain sort would be better than a blank, yet how could a logician possibly demonstrate this? So far as logic was concerned, couldn't a sceptic be right in thinking that ordinary concepts of good and bad corresponded to sheer illusions?

5. What if you accepted that the creative success of some ethical requirement was at least a logical possibility? There would then be no reason to think that its success would be any more complicated than its failure. Neither the success nor the failure would be a question of clockwork whirring, or of engineers or major-generals hatching production or annihilation plans.

6. This needn't be an attempt to get rid of a divine designer and world-creator. A. C. Ewing, possibly the greatest idealist philosopher of the twentieth century, speculated in his *Value and Reality* that the existence of a divine person could be due directly to the ethical need for such a person to exist. John Polkinghorne, too, writes that this may well be involved in the traditional notion that God is 'self-subsistent perfection' in which cause and effect come together.[3] And if the Spinozistic, pantheistic vision were wrong, so that a created world existed outside the divine being, then any causal or creative powers

[3] Ewing 1973: ch. 7; Polkinghorne 1994: 58. Here are Polkinghorne's words at greater length: 'I suspect this is what philosophical theologians are getting at in their celebrated equation of divine essence and divine existence—not just that the divine is Being with a capital B, but that God is self-subsistent perfection, identifying within himself not only cause and effect in the quality of aseity, but also supreme goodness and its instantiation.' He adds that if all this is correct, then 'extreme axiarchism (the creative effectiveness of supreme ethical requiredness) is not a Neoplatonic "Originating Principle" which might have as a consequence, in some emanating and descending chain of being, that there was "an all-powerful person, an omniscient Designer," but it is properly to be understood, purely and simply, as an insight into the divine nature itself'. Forrest (1996: 153), reasoning that it could be better 'that there be a God who creates this universe than that this universe comes into existence spontaneously', then comments that this wouldn't mean having to abandon the idea that a Value Principle was itself directly responsible for something; for 'we may now give A. C. Ewing's answer to the question, Why is there a God? namely, "Because it is good that there is a God." '

that God possessed might be explained exactly as his existence was. Granted that God was a supremely good person and therefore unable to will evil things, it might be ethically required that this person be divine not merely through knowing everything worth knowing, but through being immensely powerful as well. Saying that his existence and powers were in this way explicable—explicable, that is, by his own eternal ethical requiredness—would surely be no great insult. What praise could there be in declaring instead that the divine being was, and was powerful, for no reason at all?

7. How could anything *as intricate as* a divine person, or as an entire cosmos (pantheistic or otherwise) containing perhaps infinitely many universes, owe its existence to its ethical requiredness? I see no additional problem here. If the ethical need for a very simple thing could itself be responsible for the existence of that thing, then so could the ethical need for a more complicated thing be responsible for *its* existence. It makes no sense to add 'just so long as the thing isn't too complicated'. Remember, it is not as if somebody would have to plan the affair with the aid of complex mental machinery which would then need to be put into effect by physical machinery (cogwheels, magnets, white holes spewing out material). A modern believer in creative ethical requiredness can think as Plotinus did in the third century: that 'effort and search' play no part in the creative process (*Third Ennead*, 2nd tractate, s. 2). Suppose that my pantheistic views are right, the divine mind of which you and I are parts then being a mind that knows everything worth knowing. Did this mind have to cogitate long and hard about what would be worth knowing, beginning to think a great number of worthless things before deciding that they weren't worth thinking about any further? Did it have free choice about what to think about, so that its coming to know everything worth knowing was a matter perhaps just of good fortune or at least of choices carefully and cleverly made? Nothing of the sort occurred on my way of viewing things. The divine mind inside which we exist isn't a mind quite like yours or mine. It is just an existentially unified set of all the thoughts that are worth having, and its having them—its being them—is eternal and inevitable. It necessarily knows everything worth knowing. If, for instance, it is worth knowing that ethical requirements are what are

responsible for its existence, and that they are responsible for the existence of infinitely many other precisely similar minds, then it automatically knows these things; its awareness that the other minds exist does not depend on its entering into causal interactions with them. Or if it is good that it should contain quasi-telepathic acquaintance with all human states of mind combined with approval of some and disgust at others, or knowledge of all possible law-controlled universes, or of all possible beautiful symphonies, then it contains it, just because of this being good.

8. If an ethical need created our universe, why didn't it create it earlier? Why do only about ten billion years separate us from the Big Bang? We might answer (in imitation of Augustine) that there were no wasted years before the Bang because time and the created world came into existence simultaneously. However, Chapter 3 gave another reply. Our universe has, as Einstein said, 'a four-dimensional existence', and time is one of its dimensions. Augustine could be right inasmuch as the time dimension does not stretch backwards beyond the Bang. All the same, it makes sense to think of the four-dimensional whole as existing in time of another sort. In this other time it has always been present, and always will be. (In any talk of *creative ethical requiredness,* the word 'creative' does not imply action confined to some initial moment at which the creatively required entity came to exist. Aquinas rightly insisted that ours could be a created world whether or not it had existed for infinitely long. God's creative power would act at every moment to make the difference between it and emptiness.)

Some Objections to the Platonic Creation Story

The Platonic creation story can be attacked from many directions.

(*a*) It is often objected that ethical requirements aren't everywhere seen bringing good things into existence or saving them from destruction. I answer that nobody is claiming that every single ethical requirement has causal or creative success. Suppose pantheism is wrong. Suppose a divine person created our world outside himself. Crimes and natural disasters occur in it. If benevolent, why doesn't he intervene? The traditional reply is that any ethical need for intervention *would be overruled by other ethical needs*: for example, by the need for the world not to be one of constant miracles. Well, then: if the Platonic approach is correct, why aren't all ethical needs fulfilled? Answer: *because some overrule others*. The failure of various ethical requirements to put themselves into effect could run precisely parallel to the supposed failure of a benevolent deity to put them into effect through his omnipotent will. They could put themselves into effect only at too great a cost.

Alternatively, what if a pantheistic vision were right? What if our universe were nothing but a divine mind's having the good of knowing, in some tiny part of its thought, one of the perhaps infinitely many matters worth knowing, namely, the detailed nature of one particular possible universe, a universe having the kind of beauty and interest that come with obeying physical laws at all times? It would then follow that any ethical need for physical laws to be broken from time to time, for instance to save somebody whose parachute had failed, *would again be overruled*. Why on earth would the idea of overruling be invalidated through having Platonism in the picture?

Look, though, at how some theists approach these matters. Insisting that ethical requirements need divine support in order to become creatively influential, they protest that beautiful cathedrals, for example, don't automatically pop into existence 'as the Platonic story would lead us to expect', or that ours is a world of earthquakes and murders 'although the story says it wouldn't be'. And then these same theists are found arguing that earthquakes and murders (and no doubt also the annoyance of having to construct cathedrals) could be squared with the divine benevolence if ethical requirements *often conflicted with one another*. As if the Platonist couldn't say that ethical requirements often conflicted with other ethical requirements, so that any cosmos that the strongest compatible set of ethical requirements managed to create without divine aid could include Earth's earthquakes and its murders, and its lack of self-constructing cathedrals!

Again, theists sometimes claim that of course Plato's creation story is wrong since no noble or fine thing has ever existed just because of its being good that it should exist—as if Ewing were evidently mistaken

166

about why God exists, or as if the region of reality known to us couldn't possibly contribute anything noble and fine to pantheism's divine mind.

(*b*) 'Extreme Axiarchism' was, Ronald Hepburn notes, the name I gave in the 1970s to the Platonic theory that the goodness of any sufficiently good thing could ensure its existence without the help of any divine agent. He calls it 'a seriously defensible philosophical-religious position', one which 'stands or falls with the possibility of an objectivist account of values' (1988: 869). Here is the basis of another possible objection, then. Isn't it obvious nowadays that value is never objective?

At this point let us make no effort to refute various anti-objectivist theories dreamt up by philosophers—the 'emotivist' (or 'Boo/Hurrah') theory of good and bad; the 'prescriptivist' theory that to call something good is to tell yourself and others to favour it and all things sufficiently like it; the 'relativist' theory that, yes, things can be truly good or bad, but only *relative to* particular ways of evaluating; the 'contractarian' theory that people 'invent right and wrong' in order to 'internalize' social pressures so as to avoid having to keep doing what they don't really want to, the trick being to get yourself to loathe the idea of 'not doing the done thing', like the Englishman who can't bear to dine without his dinner-jacket.[4] Let us say simply that such theories fail to capture the ordinary senses of such words as 'good', 'bad', and 'ethically demanded'. The ordinary view is that various things are 'marked out for existence', 'called for', 'made needful or required', not by human likes and dislikes only (although those could be greatly important because the pleasure of getting what you like could itself be a great good) but through their own natures. It is a view arrived at for the unmysterious reason that people like to think they really are absolutely right in favouring this or that while their benighted opponents are absolutely wrong.

Still, the sole 'objectivity' clearly essential to the Platonic creation

---

[4] There is detailed discussion of these ethical theories in Leslie 1979: ch. 12. Leslie 1996*a*: ch. 4 goes so far as to suggest that defending them would add to the dangers now facing humankind *if* anybody listened to philosophers.

story is the objectivity of actually being out there in reality instead of being a fiction. Objectivity in the other sense, *being readily verifiable,* could be nice to have, but we could survive without it. Plato's theory of why there exists something and not nothing certainly needs the idea that anything possessing such and such characteristics (for instance, of being a pleasurable mental state, or of being a mind that knew everything worth knowing) truly would have the ethical status of being 'marked out for existence or required': required to at least some extent in the case of such things as pleasurable mental states, and required overwhelmingly in the case of a divine mind, perhaps. But why think that this would have to be readily verifiable? If it weren't verifiable at all, then so what? Hugh Rice, for example, commits himself to more than is necessary. Accepting a Platonic tale ('We could say that there is a universe such as this, obeying laws such as these, because it is good that it should be so'; an explanation which appeals 'directly to goodness' is preferable to one which introduces 'a mediating God'[5]), Rice says that the line he takes differs from the one he found in my *Value and Existence* because he sees the tale in question as offering 'an explanation of the same general sort as applies to the explanation of many beliefs we all have about good and bad' (Rice 2000: 49). In particular, he thinks that necessary facts about good and bad *can directly influence* our judgements of good and bad; they are thus facts knowable in some more or less immediate manner. Well, right though he might perhaps be in thinking this, I never have thought it. In my view, what the Good

---

[5] Rice 2000: 49 and 64. On p. 51 Rice reacts as follows to the idea of a Creator producing things by mental operations: 'the notion of a mind's creating a universe is a very unfamiliar one, and (or so it seems to me) at least as unfamiliar as the notion that the goodness of the universe is directly responsible for its existence'. He asks (p. 89) how we could possibly be helped by being told that ethical requirements, instead of being themselves effective, are put into effect by 'something of an unknown nature performing operations of an unknown nature'. See also Stephen Clark: 'What exactly is added by saying that things are so because someone wants them to be? Either there are real constraints on what that one could want or there are not: if there are not, we have no explanation that is more than a brute fact (and so no explanation); if there are, might not those self-same constraints themselves have an effect on what, physically, there is?' What does 'God's intention' add except 'an extra wheel that turns without effect'? (Clark 1990: 198 and 200). My reaction is that Rice and Clark may be overstating a good point.

of Plato has created—what exists because of its supreme ethical requiredness—is (see my 1979: ch. 11) a set of infinitely many conscious wholes of immense complexity: 'divine minds', I now call them. Our world is part of one such mind, a part which has the good of conforming throughout to laws of physics (that's what *our* divine mind contemplates *here*). Now, presumably no brain entirely ruled by laws of physics could have the kind of direct acquaintance with the facts of good and bad that Rice envisages.

Where did my beliefs about good and bad come from, then? They are products of evolutionary forces; of social pressures; of my love and respect for my parents; of the vagaries of my cerebral processes; of contempt for the argument 'If we *cannot know* what's good and what's bad, then it follows that we *can know* it's good to tolerate murder and bad to punish thieves'; of a wish to see life as intrinsically worth living.

(*c*) In *The Riddle of Existence* Nicholas Rescher suggests that the laws of nature take the form they do because of an 'axiological principle'—a principle of value, widely understood. Why does the principle hold? Answer: 'the laws are governed by a principle of what's for the best because that's for the best'. Yet isn't this to embark on an infinite series of *that's-for-the-best* explanations? Not so, Rescher replies. Instead, all that's going on is that the principle of the best 'is self-sustaining'; it 'does not require any "external" explanation at all' (Rescher 1984: ch. 2, and particularly pp. 53–4; cf. 2000: 156). (No need to introduce a divine person's power and benevolence, for instance.) Yet just what would it be about *best-ness* that could make this reply satisfactory? In 'The Puzzle of Reality' Derek Parfit introduces what he calls 'the Platonic, or Axiarchic, View'. According to it, 'that there is a best way for reality to be explains *directly* why reality is that way. If God exists, that is because His existing is best. Truths about value are, in John Leslie's phrase, *creatively effective.*' Well, best-ness is what Parfit classifies as 'a plausible Selector', that is, a factor which not only 'selects what reality is like' but is also such that 'we can reasonably believe that, were reality to have that feature, that would not merely happen to be true'. In contrast, *being fifty-seven in number* wouldn't be a plausible Selector, he writes. If exactly fifty-seven worlds existed, then their fifty-seven-ness could hardly be the reason for their existence.[6] Yet couldn't people justifiably enquire of him why best-ness would be any different from fifty-seven-ness? Why, Richard Dawkins asked me at a workshop, should 'so piffling a concept as goodness' explain the world's existence? Chanel Number Fiveness would seem quite as helpful, he said.

In replying on Plato's behalf, we can start by insisting that goodness is *required existence* in a non-trivial sense. Goodness isn't an ordinary quality like redness or pleasurableness or being perfumed with Chanel Number Five. Goodness is a status had by various possible things (pleasant mental states, perhaps), the status of being such that their existence would (and, if they do exist, does) fulfil real needs. Anybody who hasn't grasped that the good is *the required* hasn't reached square one in understanding what ethics is all about. The ideas of good and bad were formed by people who wanted to picture various things they favoured as absolutely called for, so that their opponents truly were benighted folk instead of just folk who favoured different things.

Certainly one has to be careful in understanding this. (i) Describing something as 'ethically required' doesn't mean that it, let alone absolutely all things vaguely like it, ought to be favoured on absolutely all occasions no matter what further options were available or what the causal accompaniments might be. Even calling an ethical requirement 'something absolute rather than hypothetical' needn't say that it cannot be overruled by other ethical requirements. (ii) Again, it isn't true that anyone who is ethically required to do something will therefore do it as inevitably as people stepping off diving boards will accelerate downwards. Instead,

---

[6] Parfit 1992: 3–4. Parfit is not saying he accepts the Axiarchic View. He doubts whether our world could be even a part of a cosmos whose best-ness lay in its containing all worlds that were better than nothing, for perhaps it is a world having negative intrinsic worth. Or perhaps, as on some principles of justice, the sufferings of the unfortunate could never be adequately compensated for by the happiness of others; the monistic or pantheistic view that those who suffer and those who are happy 'are, at some level, one' could be a mistake. (See also Parfit 1998. Considering whether God's benevolent intentions could help explain why reality was best, Parfit comments (22 Jan., p. 27) that 'since God's own existence could not be God's work, there could be no intentional explanation of why the whole of reality was as good as it could be'. We could reasonably conclude that 'even if God exists, the intentional explanation could not compete with the different and bolder explanation offered by the Axiarchic View'.)

limits on what acts are 'open to us ethically' are rather like limits to what we are to believe about the future, on the basis of past experience. Situations of wondering whether gravity will operate tomorrow, or whether hammers will drive nails today as they did yesterday, can have the need for particular conclusions built into them in a non-relative fashion. It isn't merely that reaching those conclusions will make you into someone standardly labelled by the word 'rational'. Rather, reaching them truly is required of you.[7] To replace 'required' by 'required *IF* you wish to be standardly describable as rational' would miss the point, because thinking in a fashion that really is required *is what being rational is.* Well, similarly with thinking and producing what you need to think and produce, ethically speaking. The idea behind talk of goodness is that when good things come into existence they fulfil requirements that aren't entirely hypothetical. Yes, somebody's enjoyment of beautiful music may have a value that is relative to whether the city is in flames; but no, its value isn't utterly relative. The enjoyment isn't simultaneously *very good* and *very bad*: very good relative to one person's ethical standards and very bad relative to those of somebody else, where neither set of standards is absolutely superior to the other. The ethically required existence of a good world is unlike the 'etiquettely required existence' of putting out your tongue as a polite greeting in Tibet or the 'thermally required existence' of a world that is red-hot. (iii) In an absence of all existing things, could it make sense to say that the coming into existence of a good world or the staying out of existence of an atrocious one, a world of great negative value, *was morally required*? Definitely not. Morality is a matter of people's duties. *No moral requirements without persons*, therefore! But equally, it would be very strange to say that there would be nothing advantageous in the continued absence of the atrocious world, nothing needful in the coming to exist of the good one; and here 'advantageous' and 'needful' are ethical words. There is no absurdity in calling the existence of good situations 'required ethically' whether or not anybody exists so as to have a duty to produce them. They are needed, called for, marked

---

[7] Rice is well aware of points of this kind. Rice 2000: 29: it is not 'simply a consequence of something which we have laid down by definitional fiat' that when a coin falls heads twenty times in a row we ought to suspect it of not being a fair coin.

out for existence, in a way for which 'ethically' is the only word readily available. (iv) Now, there is at least no conceptual confusion in the Platonic idea that the ethical requirement that there exist a good cosmos or a divine person *could itself necessitate* the existence of that cosmos or that person.

How Ethical Requirements might be Real Necessarily, and with Necessary Creative Power

Imagine some good possibility. If it were actualized, then it would enjoy an existence that was ethically needful, an existence that fulfilled an ethical requirement. The Platonic suggestion is that a thing could exist *because* its existence was in this way needful or required. Still, a thing's ethical requiredness couldn't involve its existence as a matter that was straightforwardly logical. The case, for instance, of a divine mind's existing *because this was demanded ethically* would be nothing like the case of existing as four animals because of being two sets of two cows, or of existing as a male because of being a bull.

However, we could use the idea of *synthetic necessities* here. Synthetic necessities are defined as necessities every bit as absolute as any straightforwardly logical necessity. If there are any synthetic necessities, then they must all be matters which simply couldn't be otherwise, whether or not humans are equipped to recognize them, just as all dragons must (by definition) have flame-breathing abilities. But (once again by definition) synthetic necessities would never be provable from the very definitions of words or other symbols, which is how straightforwardly logical necessities are provable. Now, there do exist some reasonably clear instances of synthetic necessity. Consider three afterimages produced by bright lights. The first is red, the second orange, and the third yellow. The red afterimage is, in colour, *necessarily* more like the orange afterimage than like the yellow one, yet this is not just a matter of what definitions have been allocated to words. Cavemen who never had a language could recognize such differing degrees of similarity. It isn't through anybody's defining orange with the words 'reddish yellow' that the yellow afterimage is more colour-similar to the orange one than to the red.

Instead it is such differing degrees of colour-similarity that make 'reddish yellow' mean what 'orange' does.

Unfortunately, as soon as people start discussing ethical requirements—long before anybody speculates that some ethical requirements might perhaps act creatively—it becomes plain that they are discussing affairs far more controversial than any colour-similarities. While we might sympathize with a man who said he 'knew for sure' that torturing babies for fun was bad, we could surely also sympathize with any philosopher who, using stronger criteria of knowing for sure, said it wasn't known for sure that anything was ever better than anything else. In fact, I get the impression that many highly intelligent philosophers consider *that nothing ever is* really better than anything else, in the sense of 'really better' that is favoured by ordinary thought and ordinary language. In two very fine books of his, *Ethics: Inventing Right and Wrong* (1977) and *The Miracle of Theism* (1982), J. L. Mackie said he himself believed that nothing was really better than anything else in that sense, which he had the courage to identify not just as the ordinary sense but as the philosophically traditional sense as well.

Mackie described goodness in the ordinary sense as too 'queer'. He was at a loss to understand its allegedly necessary linkage to normal properties. Logical linkage would not do the trick, he held, for nobody could deduce a thing's intrinsic value—as ordinarily conceived—from any non-question-begging description of that thing. Curiously, though, he combined this with two further claims. The first was that intrinsic value as ordinarily conceived *involved no contradiction*. Queer it would be, but not a logical impossibility. The second was that it *could never be added to things or taken away from them, not even by divine decrees*. Either it *just was* possessed by various things and absolutely couldn't fail to be possessed by them no matter what a divine being said or did or wanted, or else it *just was not* and absolutely couldn't be given to them by divine decrees or by anything. Which, I take it, is to say that, if in fact present, then it would be present with a necessity which although not logically demonstrable was still every bit as firm as any logically demonstrable necessity. Its presence would be something synthetically necessary. Likewise its absence, if it were in fact absent, would be a synthetically necessary absence. As a matter of synthetic necessity, the kind of intrinsic value in which people ordinarily believe would either be built into situations of certain kinds, or walled out of them. We were saddled with a synthetic necessity, one way or the other.

The curious thing is that Mackie seems to have considered that the synthetically necessary presence of such value in a state of affairs would be queer whereas its synthetically necessary *absence* would be altogether to be expected. He appears to have found nothing odd in how his theory denied that it could be, in what he himself called the ordinary sense of the word 'better', *better* to be feeding a hungry child than to be kicking one—though he would have hurried to behave in the first way, feeling fury at anyone who behaved in the second.

Mackie accepted that the ordinary idea of a situation's intrinsic value was the idea of *its having an existence ethically required in a Platonic fashion*, a fashion that was no mere reflection of social pressures. And, given that he detected no actual contradiction in this idea, although finding it too queer for his taste, it is no great surprise that he confessed in chapter 13 of *The Miracle of Theism* that he detected no contradiction, either, in the Platonic theory that an ethical requirement for a thing to exist could itself be creatively successful. So far as mere logic was concerned, it seemed to him that some possibilities might indeed, just through being the possibilities that they were and without help from any powerful person, be marked out for existence *ethically and with creative effect* as argued in my *Value and Existence*. They might be things that firmly had to exist, whether or not we could know this.[8]

Remember that it could be wrong to keep asking for some mechanism

---

[8] Mackie 1982: ch. 13, controversially entitled 'Replacements for God', its topic being the theory that ethical requirements could create a cosmos without the aid of divine acts of will. Mackie writes (p. 234) that such a theory provides 'a formidable rival to the tradition-al theism which treats God as a person or mind or spirit'. The notion 'that objective ethical requiredness is creative, that something's being valuable can in itself tend to bring that thing into existence or maintain it in existence,' is one which he calls 'pure speculation' and implausible; yet, he says, it is right to resist 'the prejudice that it can be known *a priori* to be impossible' (p. 237). He writes also (same page) that there would be no excuse 'for allow-ing the hypothesized principle of creative value to be *called* God', which would be a mere 'device for slurring over a real change in belief'; but the theologian Brian Davies told me that this showed Mackie up as knowing too little of the history of theology. (Full references to Mackie's words discussing this area are given in Leslie 1986c.)

which gave to an ethical requirement its alleged creative ability: some combination, perhaps, of pistons pushing, electromagnetic fields tugging, or persons exerting willpower so as to require with actual success that the ethically required object should come to exist. This could be like asking for a mechanism to explain how a yellow afterimage managed to be nearer in colour to an orange one than to a red, or why being in pain and hating it was intrinsically worse than hearing music and loving it. It could be a failure to appreciate how odd it might be to imagine that, say, somebody's act of will could by itself effectively require such and such, while simultaneously insisting that ethical requirements by themselves could never do so, not even in the case of the ethically required existence of a divine mind—something supremely perfect.

It could also be a failure to grasp that in this area, too, *we are saddled with a synthetic necessity, one way or the other*: either the synthetic necessity that creative effectiveness *does* characterize one or more ethical requirements, or else the synthetic necessity that it *does not*. It could not just so chance that some Platonic reality of ethical requirement had creative power, or alternatively that it was powerless. If Goodness is what Parfit calls 'a Selector' so that not only is reality maximally good but this is, as Parfit puts it, *no coincidence*, then one needs to have some way of understanding the fact of its being no coincidence; one needs some hard (or 'ontological') distinction between reality's simply happening to be maximally good and its being as it is *because* that is maximally good; and this means the presence of a necessity. The state of being Platonically required with direct creative effect seems to me the very opposite of chancing to exist. Yet neither the creative power nor the powerlessness of any Platonic ethical requirement could be provable by logicians. Mackie was right about that.

The power and the powerlessness would be equally simple. The failure of the Platonic Good to act creatively would be in no way *less complicated* than its success. It would not be 'straightforward' if an ethical requirement for a divine mind to exist remained unfulfilled, and 'unstraightforward' if it were fulfilled. The synthetic necessity we are stuck with is exactly as uncomplicated, no matter which form it takes.

175

Concessions

1. You may still want to say that an ethical requirement *in itself* and *as such* would be too abstract to produce anything. Now, in some senses of 'in itself' and 'as such', that's right. Yet maybe there is nothing here to trouble Platonists.

To protest that an ethical requirement is never a force, and therefore never a creatively effective force, amounts to declaring without argument that Plato is wrong, for according to Plato at least one ethical requirement does act as a force of a kind, a creative force. The Platonic creation story just is that creative success is had by some one ethical requirement (for there to exist a divine mind, perhaps) or compatible group of ethical requirements (for example, for there to exist all the ingredients of a good cosmos). Yet how about the argument that an ethical requirement *cannot itself do anything because it is an abstraction*? Well, this invites the reply that anything able to explain Why There Is Something Rather Than Nothing—why there is even a single existent—couldn't itself be an existent. It would have to be an abstraction of some kind. But aren't there senses in which abstractions are known to be able to do things? The fact of being square (an abstraction of a sort) can prevent a peg from fitting into a round hole with the same cross-sectional area. Again, take a disc whose two faces are almost indistinguishable. The abstract fact of their near-indistinguishability can do something when the disc is tossed a great many times. It can help make each face fall upwards more or less exactly half the time. Or consider a situation discussed in my 'Ensuring Two Bird Deaths with One Throw' (1991*a*). Suppose you lived in a universe which, fully symmetrical at its earliest moments, evolved deterministically ever afterwards so that it remained symmetrical. Couldn't the abstract fact of its symmetry help you to do things? To guarantee a disastrous end for a bird in the universe's other half you'd need only to throw a stone accurately at the corresponding bird in your half.

None the less it has to be conceded that *a creatively successful ethical requirement* would be an abstract reality composed of two still more abstract realities: its creative aspect and its ethical aspect. Now, if the creative aspect could be destroyed (impossible if synthetic necessity is in command here, yet let's imagine it for argument's sake) then couldn't the ethical aspect continue to be real? And couldn't you justifiably say that what would continue to be real would be 'strictly ethical' whereas the creative power, *because it was inessential to the ethical aspect*, 'wouldn't itself be strictly ethical' so that talk of anything's *creative ethical requiredness* would be strictly speaking wrong?

You could indeed. What you count as 'strictly this' or 'strictly that' will vary with just where you choose to cut the cake of reality when forming your abstractions. Are you inclined to say that orange isn't truly a case where *redness* (or *redness itself*) has taken on a yellow tint, because strictly speaking, in so far as something is red, it isn't yellow? Please yourself! Just don't require everybody else to cut the cake where you do.

A cow has been dyed purple. Is it purple *in itself?* That depends on whether you prefer to count 'the cow in itself' as ending before the dye begins. Maybe you'd answer in one way in any case of purple dye, in another in cases of purple paint or coatings of purple mud. Take an ordinary brown cow. Is *the cow as such* brown? You might prefer to say that every *cow as such* is female (meaning that being female enters into the definition of a cow) but not brown. Cows *qua* cows have colour but not brownness, might be your verdict. But you'd need to make clear you weren't saying that all cows are necessarily non-brown.

Can a judge *qua* judge, a judge 'acting as a judge', make private courtroom notes? With a pen? With inkblots? Judges as such sentence criminals, but these other things are surely mere matters of how people choose to speak. So long as there were *some senses* in which an ethical requirement could be creatively successful 'itself', 'as such', and 'without ceasing to be strictly ethical', then that should be enough for any Platonist. There'd be no point in denying that in other senses this wouldn't be so.

2. Similarly with whether ethical requirements 'could one and all be creatively powerless'. Conceding that there was a sense in which this was so wouldn't be a denial of the synthetic necessity discussed earlier. It would merely be recognition of a logical and epistemic possibility: the creative power couldn't be logically demonstrable or in some way obvious at a glance, and any evidence in its favour would be very controversial.

3. We might have to concede, though, that such power couldn't fail to be *in some sense* 'a matter of logical necessity'. We are in the area of synthetic necessities, remember: necessities defined as being every bit as firm as any logically demonstrable necessity. If real at all, synthetic necessities are real in all genuinely possible worlds. But today, aren't many philosophers dissatisfied with what they call 'overly linguistic' treatments of logical necessity, and don't they therefore claim that 'being logically necessary' is best understood to mean *being the case in all genuinely possible worlds*?

Also, don't other philosophers argue that so-called synthetic necessities would be seen to follow logically if only we had the right concepts? Look once more at how a yellow afterimage is nearer in colour to an orange one than to a red one. Although no mere product of defining orange as 'reddish-yellow', mightn't this be involved in any completely adequate concepts of experienced red, orange, and yellow, so that it was after all an essentially conceptual affair? And similarly in the case of creative effectiveness, perhaps. If some mind possessed concepts as adequate as anyone could wish, then those concepts might conceivably include a recognition that some ethical requirement or requirements simply had to act creatively. But I very much doubt it— unless we trivialize the affair by shovelling everything necessarily true of anything into its 'completely adequate concept'.

4. Have I established firmly even the slightest likelihood that ethical requirements ever themselves wield any power? I suspect that I have not. How, after all, could we form any estimate of anything worth calling 'the a priori probability that some ethical requirement would be responsible for its own fulfilment'? By contemplating the mechanism whereby the ethical requirement would supposedly put itself into effect, then trying to guess whether this mechanism would operate smoothly? There is no such mechanism: no complexly interacting magic spells or intricate exertions of pure willpower; no steam engines, quantum fields or anything else. So, were somebody to claim that it couldn't be proved incontrovertibly that the Platonic creation story stood any chance of being correct, then I'd say that this could well be right. I'd add, though, that it would be every bit as difficult to prove incontrovertibly that the story had any chance of being wrong. While I

myself would reply 'something like 45 per cent' if forced to estimate the likelihood that the world exists for no reason whatever, answering 'zero' wouldn't be an incontrovertibly provable mistake, I suggest.

What conditions would a requirement have to satisfy for it to stand a chance of explaining why there are any existing things? First, it would have to be a Platonic reality, eternally and necessarily real. It could not depend for its reality on there already existing some powerful person or thing: a self-moving magic wand, for instance. Second, it would have to involve *existence being demanded not just hypothetically*—in contrast to the reality that *if* there existed three sets of five tame tigers, then this would require there to be fifteen, or that *if* you want to annoy the neighbours then you ought (in a non-ethical sense) to play loud music at all hours. Third, its creative success couldn't be logically demonstrable because not even a divine mind can have a logically demonstrable existence. Well, my claim is that ethical requirements satisfy these three conditions and that there is no contradiction in the idea of an ethical requirement that itself acts creatively. Do you still think nothing has been done to show that some ethical requirement could have acted creatively, with a probability above zero? Your wrongness may not be firmly demonstrable. Yet you might be judged rather too like somebody who learns that the thief has green eyes, an orange beard, and a drunkard's nose, and also that Mr Jones has all these things, but who views this as suggesting nothing.

Competing Pictures of God

If a Platonic creation story were right, then how would God best be pictured?

(i) God as an abstract force or principle, or a creative aspect of the cosmos

We might conceivably prefer the line I said had been taken recently by Rice. In his *God and Goodness* Rice defends 'an abstract conception of God' according to which the statement that God created the world says merely that the world exists 'because it is good that it should'; there is no question of God as 'a concrete *something* which wills and creates'. On the abstract conception, 'God is not a *person*' (2000: 88). Rice argues, however, that the principle that the world *owes its existence directly to the ethical need for it* yields very much the results that a benevolent person would. The Platonic theme met with here, that the Good can be influential in producing something without a divine agent's aid, had (as I noted on earlier pages and in the footnotes to them) strong attractions for Nicholas Rescher and (see n. 5) for Stephen Clark, also receiving polite treatment from Ronald Hepburn, Derek Parfit, and even J. L. Mackie, a militant atheist. The theme can be combined (as in Rice's book) with rejection of the idea that an omnipotent designer would be an entity which creative value would hasten to create, and the combination has a distinguished ancestry. In book 1 of *The Divine Names*, Pseudo-Dionysius held that *what is not* is the cause of all that is, likening the divine reality to 'our sun which, without choosing or taking thought, enlightens all things'. The First, Third, and Sixth of Plotinus's *Enneads* tell us that the Good 'on which all else depends' is not itself an existing thing for it has 'no need of Being'; our world exists 'not as a result of a judgement recognizing its desirability, but by sheer necessity'; yet the outcome, 'even had it resulted from a considered plan, would not have disgraced its maker'. And people have held that Aquinas thought similarly. In a recent issue of *The Monist,* Hilary Putnam (1997) presents himself as an 'Analytic Maimonidean'. Attracted by the idea that traditional arguments for God's reality point towards 'necessities which are not simply conceptual', he inclines towards what he takes to be Aquinas's view that God is not good or wise or a being in the sense in which created persons are good, wise, and beings. God is instead 'the "principle" or ground of what we call goodness and wisdom and being in creatures'—this corresponding exactly, Putnam says, to the account Maimonides gives of the 'attributes of action' that we are permitted to ascribe to God. Another article in the same issue, by Brian Davies, quotes Aquinas's statement that God is 'outside the realm of existents, as a cause which pours forth every thing that exists'.[9]

Leibniz might perhaps be counted as reasoning similarly, particularly in view of his *On the Ultimate Origin of Things*. Sketching a creative process in which the tendencies towards existence of various competing goods resemble conflicting pulls of heavy bodies, Leibniz comments that 'from the fact that there exists something rather than nothing it follows that in possibility or essence itself there is a need of existence, a claim to exist'. Bertrand Russell saw remarks of that kind as pointing towards a secret Leibnizian philosophy in which there was no mention of any divine person. Still, it seems to me that Leibniz's God really was a person: one who was far the best member of a cosmos that existed as a direct consequence of its ethical requiredness.[10] In contrast, actual denials of God's personhood have been made by such more recent philosophers as Henri Bergson and A. N. Whitehead and by such modern theologians as Paul Tillich, J. A. T. Robinson, and Hans Küng. In part 3 of *The Two Sources of Morality and Religion,* his poetically exuberant work (1935), Bergson identifies God as 'creative energy'. ('Beings have been called into existence who were destined to love and be loved, since creative energy is love. Distinct from God, who is this energy, they could exist only in a universe, and therefore the universe sprang into being.') In Whitehead's *Religion in the Making* God is part of 'the realm of ideal entities, or forms, which are not actual but are exemplified in everything actual'; the actual world is 'the outcome of the aesthetic order', which is in turn 'derived from the immanence of God'. Tillich's *Systematic Theology* actually calls it 'blasphemy' to make God 'one being among others'. If God were only a

---

[9] Putnam 1997: esp. 487, 489, and 496. Davies 1997: 517, quoting from Aquinas's *Commentary on Aristotle's 'Peri Hermeneias'*. Compare *Summa Theologiae*, Ia, qu. 5, art. 2, 'Goodness as a cause is prior to being', or the statement in *Contra Gentiles*, 3.20, that 'even non-existent things seek a good, namely to exist'. Ivor Leclerc remarks that Aquinas interpreted *esse* 'as fundamentally a verb and not a noun' so that God, identified as *ipsum esse* (being itself, very being), could be *actus purus* (pure act). And Aquinas was correct in Leclerc's opinion. God 'is not to be conceived as "a being", but as the "principle or source" of being' (Leclerc 1984: 73 and 78). We must follow 'Plato and Plotinus' in maintaining '*not* that God is the *principal* good, but that God transcends good by being the *principle of* good' (Leclerc 1981: 25).

[10] This way of reading Leibniz is discussed at Leslie 1979: 200–4.

highest being, then he would necessarily have a cause beyond himself, Tillich claims, so 'to argue that God exists is to deny him'; God is instead 'the creative ground of existence'. In Robinson's eyes 'theological statements' do not describe any 'highest Being', as is understood by 'the classical Platonist tradition of Christian ontology'. And in Küng's *Does God Exist?* we read that God 'is not a supramundane being'; 'he is not an existent', as has been recognized by writers stretching 'from Pseudo-Dionysius to Heidegger'.[11]

(ii) God as a person existing for reasons of value, and perhaps creating everything else

I cited John Polkinghorne and (in n. 3) Peter Forrest as attracted by A. C. Ewing's theory that God is a person owing his existence to an eternal ethical requirement. Again, when noting that Keith Ward thought that something's supreme goodness might bear responsibility for its existence, I could have added that he viewed a divine person as the 'something'.

God-as-person would strike many people as the sole entity whose existence could be accounted for in this way. Any ethical requirement that bore responsibility for a divine person's existence could bear responsibility for his omnipotence as well, and it would then be up to him to decide whether anything else would exist. Forrest, remember, goes so far as to suggest that it would be better for the existence of our universe to be due to a person's creative act. ('Why does not what is good just come about anyway?', he asks (1996: 153). Perhaps because 'the production of what is good by someone acting for reasons is itself better than the spontaneous coming to exist of what is good'.) On the other hand, we might prefer the picture which I think of as Leibnizian. God could be simply the central, supremely good element in a cosmos all of whose elements existed thanks to the ethical requiredness of the whole. All other beings in the cosmic system might owe their place in

[11] Whitehead 1927: ch. 3; Tillich 1953–63: i, chs. 7 to 8; Robinson's best-selling *Honest to God* (1963: ch. 3) and his contribution to *The Honest to God Debate*, a volume edited by D. L. Edwards; Küng 1980: 185 and 601–2.

this system—might owe the fact that any ethical need for them to exist wasn't overruled by other ethical needs—almost entirely to their relationship with its central element. This would give God immense *creative influence*.

Ewing's theory is not radically new. Ewing views himself as only spelling out what a great many religious thinkers have had in mind when they suggested a link between God's existence and God's goodness, a connection sometimes wrongly presented as logically demonstrable in the style of the so-called 'Ontological Proof'. (Perhaps we should say Ontological *Proofs*. Anselm maintains not only that *existence* is part of the perfection which God possesses by definition, but also that *necessary existence* is part of this perfection.) The theory perhaps originated in Aristotle's remark that the First Mover exists necessarily 'and inasmuch as he exists by necessity his mode of being is good' (see n. 2).

Is such a position an attack on God's glory? If it made God's existence depend on something outside God, this still mightn't be considered disparaging. After all, God's own possibility cannot have been under God's control, and the logical, mathematical, and ethical truths which God supposedly knows wouldn't be under his control either. God couldn't have made misery into something very good in itself. God couldn't have made two and two make five. But in any case, 'something outside God's control' wouldn't necessarily mean something *outside God*. There is only a verbal difference between declaring that God owes his existence to an ethical requirement that he exist, and declaring instead that he owes his existence *to his own ethical requiredness*.

In his *First Reply to Objections* Descartes remarked that, although God had never been non-existent, he could be named 'the cause of his existence' since his existence followed from his essence. Not all *following* has to be logically demonstrable following and, as the *First Reply* rather indignantly insists, Descartes had been aware of this when running his Ontological Argument: 'a word's implying something is no reason for that thing's truth', he comments. When Ewing writes (1973: 157) that God's existence 'will be necessary not because there would be any internal self-contradiction in denying it but because it was supremely good that God should exist', and that 'the hypothesis that complete perfection does constitute an adequate ground for existence does seem to be the only one which could make the universe intelligible', he can be looked on as following in the footsteps of Descartes and of numerous others who have described God as *causa sui*—'self-caused'—or as 'existing through his own nature'. As an affair not of logically demonstrable necessity but of synthetic necessity, a supremely good possibility could have a character which called for the actualization of that possibility *ethically and with creative effect*. If the possibility in question were a divine person, then this person's own nature could require his existence with omni-temporal or eternal success. Always bear in mind that in theology words like 'creation' need not refer to the bringing into existence of something where previously there was only emptiness.

In short, a Platonic creation story is fully compatible with Ward's statement (1996a: 195) that 'nothing other than God can account for God'. 'Either', Ward continues, 'God cannot be accounted for—which makes the divine existence and nature something which just happens to be the case—or the divine nature can account for its own existence' so that God would be 'a self-explanatory being'. And to grasp what would be going on there, one seemingly needs to understand that God's ethical requiredness would be involved. 'This', Ward says, 'is the best of reasons for the existence of a being of supreme goodness, namely, that its existence is supremely desirable, not least to itself.'

(iii) A pantheistic divine mind, or an infinite collection of such minds

Spinoza's God is an all-inclusive divine mind existing because this is best. 'Whatever is, is in God' (*Ethics*, Part One, Proposition Fifteen), and God is perfect in a sense which doesn't make nonsense of Spinoza's title, *Ethics*. True, God does not work for ends, but this is because if he did then he would be seeking 'something of which he stands in need' whereas his perfection means that he lacks nothing (*Ethics*, Appendix to Part One); being perfect, he 'cannot change into anything better' (*Short Treatise*, Book One, Chapter One). While the divine intellect is *natura naturata* (*Ethics*, Part One, Proof to Proposition Thirty-One), the principle whereby it exists, *natura naturans,* is clearly a principle of the good: 'God has from himself an absolutely infinite power of existence', for 'perfection does not prevent the existence of a thing but establishes it' (*Ethics*, Part One, Note to Proposition Eleven) so that 'through his perfection, God is the cause of himself' (*Short Treatise*, Book One, Chapter Three).

Similar themes appear in Hegel's writings. 'The universality moulded by Reason, the final end or the Good', is something which should be viewed 'as actualized', and actualized thanks to 'the power which proposes this End as well as actualizes it,—that is, God' (section 59 of the first part, on Logic, of Hegel's *Encyclopaedia of the Philosophical Sciences*); the Idea which thinks itself, supreme in its goodness, 'is not so impotent as merely to have a right or an obligation to exist without actually existing' (section 6). British Hegelians such as F. H. Bradley, J. M. E. McTaggart, Bernard Bosanquet, and A. E. Taylor concluded that reality was so perfect that no one ethical requirement ever ultimately overruled another. 'Heaven's Design', as Bradley put it, 'can realize itself as effectively in "Catiline or Borgia" as in the scrupulous' because the timeless perfection of Absolute Reality 'stands above the element of event, contradiction and finitude' (1893: ch. 17; and 1927: Concluding Remarks). This strikes me as a thoroughly unacceptable doctrine; but if one keeps well away from it, then a Spinozistic account of the divine nature has much to recommend it.

I have been arguing, though, that a scheme of things which existed because of its ethical requiredness would be far richer than the one Spinoza described. There would be infinitely many immensely knowledgeable minds, each contemplating the details of innumerable universes.

Competing Senses of 'God', 'Platonic', and 'Neoplatonic'

Perhaps the reason why there exists something, not nothing, does lie in an eternal ethical requirement for the existence of a cosmos which is immensely good. We might then want to give the name 'God' to the principle that ethical requirements are creatively effective when not overruled by other ethical requirements. Or we might give the name to the creatively effective ethical requirements themselves, considered as joining compatibly to form a single requirement: the creatively effective need for the existence of an immensely good cosmos. Or again, we could take the Spinozistic line that the cosmos is worth calling 'God' *because* it exists in fulfilment of such a requirement. Or we might use 'God' as a name for an aspect of the cosmos: *its creative ethical requiredness.*

The competition between these various uses for God-talk is of precious little interest because *the situation actually believed in* is the same in each instance, isn't it? Denying 'that everything is God', Tillich nevertheless tells us that God is 'the power of being in everything', a 'creative ground of existence' which is also 'the foundation of moral principles' (1953–63: i, chs. 8 to 10). In plainer language, his belief is that God is *the creative ethical requiredness of the cosmos in its entirety.* Well, what difference is there in what you actually believe when you say instead *that God is the cosmos in its entirety, conceived as having creative ethical requiredness?* There is none that I can see. It is just that the word 'God' is being used differently in the two cases.

What if we believe that the cosmos—Absolutely Everything in Existence—contains an infinite number of vastly knowledgeable minds and nothing else? We might attach the label 'God' to the cosmos as a whole. Alternatively, we might take 'God' to mean the vastly knowledgeable mind inside which we ourselves exist, speaking of the other minds as 'other deities'. Or we could take 'God' as a word for the creative ethical requirement that was responsible for the whole, or as naming the creative ethical requiredness of the whole. Once more, the various options differ only verbally. The believed-in situation is in each case the same.

Suppose, however, that we think there exists an immensely impressive person who created everything else. We might well want to use 'God' as a name for this person. In fact, we might want to use 'God' to mean the immensely impressive person even if we thought of him not as *the Creator* but just as a centre around which everything else was organized—which, as I said when discussing Leibniz's position, could give him vast *creative influence* inside a cosmos whose every component existed as a direct consequence of the ethical need for the whole into which it entered.

Notice, though, that a philosopher who believed that the cosmos as a whole existed as a direct consequence of the ethical need for it might believe in the impressive person yet prefer not to call him 'God'. For this philosopher, 'God' could name the principle that the cosmos—the impressive person plus everything else—exists just because of an ethical requirement. In his book called *God and Goodness* (like Rice's) Mark Wynn tackles this point (1999: 66). Leaning towards the view that 'God' is best treated as the title of 'a set of causally efficacious ethical requirements' rather than of a person, Wynn remarks none the less that 'if we grant that consciousness is a profound value, then we are likely to suppose that there is an ethical requirement that there be a supreme consciousness', in which case the title may be one to which the set of causally efficacious requirements and the supreme consciousness 'both have a claim'.

Note, too, that somebody accepting Spinoza's world-picture might choose not to follow Spinoza in applying the name 'God' to the world as a whole. Remember, Spinoza appears to accept that, in addition to all the various divine thoughts which are the various things in our universe, there is a 'divine overview' in which everything is grasped as if in a single glance. Furthermore he seems to have imagined this overview as endowed with personality of a sort. Besides loving himself because 'the nature of God delights in infinite perfection', Spinoza's God loves individual humans, his love for them entering into 'the love with which he loves himself'—not, in this case, 'in so far as he is infinite', but 'in so far as he can be manifested through the essence of the human mind' (*Ethics*, Part Five, Proposition Thirty-Five and its Proof, and Proposition Thirty-Six and its Corollary). Well, might you not reserve the word 'God' for this divine overview and personality?

Similar ambiguities attend the terms 'Platonic' and 'Neoplatonic'. You might use them interchangeably, saying that a 'Platonic or Neoplatonic story' is any tale that answers Why There Exists Something Rather Than Nothing by pointing to a creatively effective ethical requirement: a requirement perhaps just for the existence of a divine person. But alternatively, with an eye on the importance of Plotinus, you might want to reserve the word 'Neoplatonic' for the story that absolutely everything owes its existence to a creatively effective ethical requirement. If any such requirement created only a divine person, it being left up to him whether anything else got created, then that's Platonism but not Neoplatonism, you might state. Or again, you might say 'Neoplatonic' only of a creation story that positively denied the existence of a divine person.

Of these various verbal habits, none is any more right than any other. Choose between them as you please!

The Evidence 6

For reasons examined earlier, we might seem to have no evidence against a Platonic creative principle. Reality might consist of infinitely many divine minds each infinite in its richness, which would make it immensely good. True, it might be considered preferable for any divine mind to be ignorant of miseries such as you and I experience, yet we might instead think that such ignorance would be bad. Also the limitations to our experiences seemingly cannot refute the theory that all experiences occur inside divine minds, because such a theory can itself maintain that divine minds know precisely how it feels to have limited experiences. Still, what data could bring positive support to Plato's theory that the Good has acted creatively?

For a start, the sheer fact that a world exists ought not to be overlooked. It may be impossible to prove that there is anything problematic here. At the same time, people are not being silly when they are puzzled by it—puzzled at there existing more than just a realm of facts about possibilities. Pointing out that if nobody actually existed, then nobody would be there to be puzzled, is a poor argument against their puzzlement. And one cannot get rid of it by declaring that the world has existed for infinitely long, 'which removes the problem', or alternatively that time itself did not exist before the Big Bang, a Bang that was 'quite to be expected' as a quantum vacuum fluctuation or something of that ilk. The fact of a world's existence can therefore seem a major item of evidence so long as there are no strong competitors for a Platonic approach to explaining it. Now, it could well seem there were no competitors at all, once one refused to accept that a Creator simply happened to exist.

Our world's causal orderliness, too, might be counted as evidence. After all, Platonism can offer to explain this; it is hard to see how else it could be explained; and treating it as an entirely reasonless fact can be thought highly unsatisfactory.

Again, the existence of intelligent life could be evidence. While Darwinian explanations are correct, they cannot themselves say why our cosmic environment is of a type in which Darwinian processes can lead to things as remarkable as birds, whales, and human beings. The apparent 'fine tuning' of our universe—the fact, that is to say, that many of its basic features seem such that very slight changes in them would have made it impossible for intelligent life to evolve—might best be explained Platonically. (A theory sometimes thought to compete with the Platonic one runs as follows. There exist many universes with varying characteristics. Ours is one of the perhaps very rare universes with properties favouring the evolution of intelligent living beings. But such a theory could well seem inadequate if standing all alone. It faces the problem that the strength of a physical force or the mass of an elementary particle often appears to need tuning to a particular figure for several different reasons at once. How could one and the same figure manage to satisfy several different requirements? The matter could readily be treated as an observational selection effect—something falling under the 'anthropic principle' that observers must always find themselves in life-permitting situations—only, it seems, if fundamental laws themselves varied from universe to universe. Well, how could this be plausible anywhere outside the extremely rich situation to which Plato's creative theory points?)

Possible Evidence in Favour of a Pantheistic and Platonic Account of the World

Suppose you grant that this book's Platonic theory about why the world exists isn't in clear conflict with our actual experience of the world, in particular because its approach to the theological Problem of Evil is neither silly nor wicked. Might the theory still be too much like a belief in Father Christmas through having no evidence in its favour? While this final chapter will suggest that, no, there is quite a lot of evidence for it, let it be clear that all such evidence is thoroughly controversial. Many reasonable folk will be unimpressed by it. All the same, they would have little right to dismiss it as 'not evidence at all', or at least (to phrase the point less aggressively) they would be wrong if claiming that nobody rational could count it as evidence.

The evidence is of three kinds. First, there is the fact that there exists 'something rather than nothing', something beyond facts about mere possibilities. Second, there is our universe's causal orderliness. And third there is the appearance of cosmic 'fine tuning for life', the seeming truth that if our universe's properties had been slightly different in any of quite a large number of ways then living things (or at any rate ones complicated enough to be conscious and intelligent) would never have been able to evolve in it.

The Evidence Provided by the World's Existence

Some have denied that the idea of an empty cosmos—a situation inhabited only by facts about possibilities—makes any sense. (*a*) Remember the modal realism of David Lewis. According to this, absolutely all possible things have to exist somewhere since the difference between *being actual* on the one hand, *being merely possible* on the other, is on a par with that between *existing here, in London,* and *existing over there, in Paris.* To Lewis, all so-called mere possibilities are 'over there' in worlds existing parallel to ours. While incredulous stares can scarcely refute him, his position has appealed to only few people and (see Chapter 1) it could appear to throw severe doubt on our practice of taking the past as a guide to the future, for the range of possible worlds that were orderly only up to a given stage could be thought immensely larger than the range of those that were orderly from start to finish. (*b*) Next there is the theory defended at times by such philosophers as F. H. Bradley and Henri Bergson, that an empty universe cannot be conceived and is therefore impossible. It is argued that the reality of emptiness would be a contradiction because it would involve somebody's actually existing to think about the emptiness, or else because the idea of any one thing as being absent is always the idea of something else being present instead. However, most philosophers consider these lines of reasoning faulty. They would seem to be efforts to show that the existence of something-or-other is among the matters necessary in a logically demonstrable manner. But unfortunately for that project, logically demonstrable necessities are all of them merely IFy-THENy, as in 'If there exist four wives then there are at least three women'. Such necessities can never themselves account for the actual existence of anything.

Let us assume, therefore, that the notion *that there might have been no existing things* does make good sense. Must a thing's actual existence then always present a problem? Many people see nothing problematic here. There had either to be no existing things, or else one or more of them. The second of these alternatives happens to be the case, they say. Why should it be thought any more in need of explanation than the first would have been, if it had been the case instead? How, they inquire, could experience tell us about a general tendency towards nothingness, a principle that no entity ever exists without some sufficient reason? They are unimpressed by Leibniz's protest in his *Principles of Nature and of Grace* (s. 7) that nothingness would be simpler than somethingness. They deny that it would be simpler, or else inquire 'What's wrong with complexity?'. Attempts to change their minds cannot, I think, appeal to principles that are obviously correct. Either you have a hunch that the mere existence of a thing would always need explanation or, like them, you do not. Asked 'Do you quite expect suddenly to find yourselves with extra arms and legs?', they can reply 'No, for arms and legs don't suddenly materialize in our universe; but some things do, namely, particles that spring into existence as quantum fluctuations; and other universes might leap into being for no reason at all.' This concludes the discussion—unless, that is, they consider it forceful to add that there were infinitely many possible ways of there being something, as against only one way of there being nothing, then suggesting that this made nothingness *infinitely unlikely*.

It strikes me, though, that the sheer existence of something rather than nothing *does* call for explanation (yes, no matter how many ways there were in which there might have been something). That's my hunch, which may strike you as no worse than the hunch that there is nothing calling for explanation here. And, thanks to the labours of Hume, Kant, and R. G. Collingwood, interpreting the world success-fully now seems to many philosophers inevitably dependent on vari-ous hunches that cannot find their justification in anything beyond themselves. They are not products of experience. To an organism rely-ing solely on experience, the world would for ever remain a blooming, buzzing confusion.

Well, then: what if the sheer existence of something rather than nothing did call for explanation? What viable theory or theories could explain it? In Chapter 5 I suggested that a Platonic theory could do the job. Has it any competitors?

1. How ought we to react when people say that the explanation for the existence of our universe at the present moment is that it existed at an earlier moment, and that the same applies to its existence at this earlier moment, and so on ad infinitum? Some physicists do believe in a cosmos that has existed for infinitely many years. The Big Bang, they think, was preceded by an infinitely prolonged contraction or by an infinite series of cosmic oscillations (Bang, Crunch, Bang, etc.) or was perhaps just an incident in the career of an eternally expanding reality that constantly gives rise to more and more Big Bang univers-es. And other physicists consider that our universe, while its past is finite when measured in years, has existed for infinite time on some other reasonable measure (for instance, according to imaginary clocks which ticked ever faster at ever earlier, ever more violent moments in the Bang). Yet such scenarios can deserve a Leibnizian reaction. The existence even of an infinite series of past events couldn't be made self-explaining through each event being explained by an earlier one. If—an example used by Leibniz—a book about geometry owes its pattern to copying from an earlier book about geometry, and this in turn to copying from a yet earlier one, and so on ad infinitum, then that still leaves us with no adequate answer to why the book is about geometry. The entire series of geometry books remains in need

of an explanation. (Think of a time machine that travels into the past so that nobody need ever have designed and manufactured it. Its existence forms a self-explaining temporal loop! Even if time travel made sense, this would surely be nonsense.)

2. Elementary particles (protons, for instance) can 'quantum-fluctuate' into existence through taking advantage of the Heisenberg uncertainty relationship between energy and time. They can 'borrow' the energy needed for them to exist, so long as they then 'pay it back' sufficiently quickly by vanishing. The more massive a particle (and so the more energy represented by its existence, in accordance with $e = mc^2$), the faster the loan must be repaid, yet in 1973, in the journal *Nature,* Edward Tryon suggested that our entire billions-of-years-old universe might be a fluctuation in the vacuum of 'some larger space', managing to exist for so long because no appreciable energy borrowing had been required. Tryon noted that the universe's gravitational binding energy enters the physicist's equations *as a negative quantity.* (In this it is like all other binding energies. The energy that stops a deuteron from falling apart, for example, makes it *less massive,* which is crucial to how stars burn.) It is often argued that the total energy of our universe must in consequence be exactly zero, if 'total energy of our universe' is a phrase that makes sense. Cosmologists such as Jim Hartle, Stephen Hawking, and Alex Vilenkin have developed this theme in ways that do not make use of Tryon's 'larger space' (Hartle and Hawking 1983; Hawking 1988; Vilenkin 1982). Our universe has even been described as having perhaps quantum-fluctuated into existence 'from nothing'—and while at times the 'nothing' turns out to be a chaotic space-time foam with fantastically high energy-density, at other times 'literally nothing' is said and apparently meant. Again, it is sometimes claimed that this or that quantum-theoretical model of creation has got rid of the only location at which a divine creator could appear to have been needed, namely, some first moment of cosmic time. It is suggested, for instance, that time becomes more and more space-like at earlier and earlier moments in the Big Bang, or that our universe sprang from a point which was 'outside space and time entirely' (and not in a 'space–time foam' either) but which was still somehow governed by the laws of quantum physics. Actual calculations have been made, purporting to show how likely our universe had been to quantum-fluctuate into

existence 'from nothing' or from such a point. When, though, we are try-ing to answer Why There Is Something, Not Nothing, these various manœuvres might be thought to have contributed nothing relevant.

Why? Notice for a start that philosophers and theologians, whether or not joining Augustine in thinking that God 'created time and the world together' many years ago, have only rarely held that divine creative activ-ity would be confined to bringing the universe into existence at some par-ticular moment. Aquinas, remember, wrote that the universe would have been created by God even if it had existed for ever. Again, Descartes was merely following scores of others when he held that the universe would immediately vanish were it not for God's action of 'conserving' its exis-tence. Well, the basic point that inspired Aquinas and Descartes might nowadays be expressed as follows. No matter how you describe the uni-verse—as having existed for ever, or as having originated from a point outside space-time or else in space but not in time, or as starting off so quantum-fuzzily that there was no definite point at which it started, or as having a total energy that is zero—the people who see a problem in the sheer existence of Something Rather Than Nothing will be little inclined to agree that the problem has been solved. Writing down an equation for 'the probability of something coming from nothing', even if it made some weird kind of sense, would still leave us with the question of why the equation *applied to reality*. (It would be no use saying that the equation was a quantum-physical one which itself told us it had such and such a probability of applying to something real. The question would remain of why the equation *was right*.) Hawking, having suggested in *A Brief History of Time* that his mathematics for describing the birth of our universe left no place for a Creator, came to concede that one could well need some creative factor to 'breathe fire into the equations'.[1]

3. Might it help if we were told that There Existing Something And Not Nothing is a prerequisite of anybody's wondering whether to be puz-zled at anything, so couldn't itself be a respectable source of puzzlement? Surely not. ('We mustn't be puzzled by what's necessary to anyone's

---

[1] Hawking 1988: ch. 8 (particularly its closing words, 'What place, then, for a cre-ator?') and Hawking 1993: ch. 9, last paragraph ('What is it that breathes fire into the equa-tions?'; 'Why does the universe bother to exist?').

asking whether to be puzzled' can be seen to be silly when you inspect it closely. If a gigantic comet hits the Earth but humans still find themselves alive afterwards, they won't be able to say to themselves that this could be puzzling only if there were extraterrestrials to be puzzled by it.)

4. How about saying that God, being infinite, is in a way immensely simple and also such as couldn't have been created by anything outside God? Would this mean that the divine existence, at any rate, needed no explanation? I cannot see it. Even granted that an infinite being would be immensely simple, which may be a great deal to grant, it could still be thought that there would have been less of a problem if nothing at all had existed. Richard Swinburne's infinite God as the ultimate brute fact on which all else depends is still only a brute fact. ('Not being dependent for its existence on anything beyond itself' may be a sense theologians have sometimes given to the words 'necessarily existent', but this strikes me as unfortunate. It would be so easy for laymen to get the impression that brute facts were being denied.)

5. What about the idea that God is Pure Being, with existence as his only attribute? I doubt whether this makes any sense; but even if it did, all that would seem to follow would be that *if God existed* then existence would be God's only attribute, and not *that God existed necessarily*. Much the same goes for the view that God is by definition absolutely perfect and that being necessarily existent would be part of God's perfection. The most this could conceivably show, I think, is that *if* necessary existence were indeed possible, then God would possess it. Now, while a Platonic approach may throw light on how there could be such a reality as necessary existence, I see no other means of throwing light on this.

It may be protested that we shouldn't allow ourselves to wallow in something as speculative as the Platonic theory that ethical requirements can sometimes act creatively. Yet since when has it been *nonspeculative* to maintain instead that a divine being or the universe exists for no reason whatever?[2]

---

[2] For more on the themes of this section, see my; 1978*b* Parfit 1992, 1998; and Grover 1998; or for some vigorously different opinions see Grünbaum 2000. On quantum-theoretical models of the creative process, see Halliwell 1992.

Causal Orderliness as Further Evidence

Our world's events, although certainly complicated, have the kinds of comparative simplicity that allow us to predict the future to some extent and to discover scientific laws of various degrees of firmness: for example, the law of conservation of energy whose precision keeps ruining perpetual motion machines. Despite the difficulties of saying exactly what ought to be meant by it, talk of 'laws of cause and effect' isn't just hot air. Now, does this constitute a problem? Might our world's causal orderliness need to be explained?

Half way through the last century it seemed to many philosophers that causal orderliness in general couldn't possibly be explained. Explanation of any one causal law, for instance that water boils when heated sufficiently, would, they said, have to appeal to some other, more fundamental causal law, such as the law that faster-moving molecules can more easily break free from one another. Explaining absolutely all causal laws was a ridiculous project, they concluded. They further reasoned that there could be nothing 'a priori improbable' in a reign of law rather than of disorder. After all, it can be extremely difficult to guarantee disorder of the kind that casinos like, disorder 'reflecting the distribution of possibilities'. Roulette wheels have to be constructed very carefully if they are to generate *reds* exactly as often as *blacks,* thereby allowing casinos to avoid bankruptcy. Inspired by this, A. J. Ayer actually argued that 'doing better than chance' in card-guessing experiments could by itself prove nothing. The only thing, he wrote, that could be remarkable in such experiments would be that some individuals were 'consistently rather better at guessing cards than the ordinary run of people' (1970: ch. 7). If everyone just did know which cards were which without looking at them, then their knowledge would be unmysterious. Suppose that, to win a billion dollars, you must guess the natures of twenty successive cards without making more than five mistakes. A mathematician tells you that far fewer than one in a million of the possible sets of guesses would make your fortune, but so what? Why should anything as abstract as 'the distribution of possibilities' have any influence on the world?

Later philosophers have tended to find such reasoning wrongheaded. It is usually ascribed to Hume, who is typically pictured as firm in defence of a Regularity Theory of Causation according to which causal patterns are affairs of *what simply does happen to be succeeded by what* in ways that can be relied on. In fact, however, Hume can be found suggesting something different: that not only have nails regularly moved when hammers hit them, but also *they wouldn't have moved* had the hammers missed—assuming of course that no rocks or musket balls had hit them instead. In his *Enquiry Concerning Human Understanding* Hume wrote (in s. 7) that a cause could be defined as 'an object followed by another, and where all the objects similar to the first, are followed by objects similar to the second', it here being understood that *if the first object had not been, the second had never existed*. Many people nowadays think this more or less correct, from which they conclude that the causal patterns of the past *were not* just matters of what regularly did happen, with nothing further lurking in the background to bear responsibility for their happening thus and not otherwise.

Here is another way of looking at the affair. Why does our world bear patterns such as led E. P. Wigner (1960) to write of 'the unreasonable effectiveness of mathematics in the natural sciences', or James Jeans (1930) to talk of a mathematically minded creator? Sure enough, no world could possibly disobey mathematical principles; three fives could never make sixteen; but why do event-sequences so often *illustrate* fairly simple mathematical laws? Why do so many things follow elegantly elliptical paths so that the mathematics of ellipses becomes useful?

It is always in theory possible to view events as having obeyed vastly complex laws rather than simple ones. There are countless mathematical formulas able to describe a line even if it has been a straight line for miles. (A line which had run straight, straight, straight, across an entire galaxy, might be just the beginning of one that then wriggled wildly, and countless astronomically lengthy equations could in principle be found that would generate lines with such initial straightness and subsequent wriggliness.) However, fairly simple formulas and not immensely long ones are what scientists see as *really* ruling the world, this being what encourages them to make the predictions that they do. If you believed astronomically lengthy equations had in fact been in charge, then you'd have to think all the history books entirely mistaken about what had caused what: every bit as mistaken as if the world had in fact been governed by pure chance, like an encyclopaedia whose pages had been generated by a monkey typing randomly. In effect, *why causal orderliness should be seen as a problem* appears to be closely linked to *why causation as ordinarily understood should be viewed as real at all.* This is among the things that make one think Hume and Kant and Collingwood were in pursuit of something important. We must view past events as having been somehow forced to obey comparatively simple laws. Other ways of viewing them, for instance as products of pure chance, *are wrong*.

You can be impressed by causal orderliness without claiming that it was 'a priori improbable' in the sense that an intelligent being without actual experience of the world would have predicted it wouldn't be found in it. The point is instead that, once having found it, we might well think it needed explanation. Had the world turned out to be entirely composed of perfect cubes of various sizes, could it have been sensible to say, 'Well, that's how things just happen to be'? (What if the Greeks had said to themselves that triangles just did happen to have angles always adding up to the same thing? Causation isn't a matter of mathematical necessities like those on which Euclid threw light; but the point is that when you find a regularity then the search for an explanation makes sense, doesn't it? Roulette wheels may have to be superbly engineered if casinos are to stay in business, yet what would you think of a wheel whose every three successive *reds* were followed by three successive *blacks*? You don't need superb engineering to avoid *that*.)

It could be a mistake, however, to assume that why elementary particles perform their mathematically elegant pirouettes calls for one answer whereas why any particles *exist* calls for another, entirely distinct. Why not roll the two questions into one, asking why there exists a world like this, a world of particles moving in these elegant ways? For Platonists like me, at any rate, the presence of Something Rather

Than Nothing and the structure of the world's events can be accounted for in the same manner. If this leads naturally to pantheism, then
one's explanation of why the world exhibits causal patterns will tend
to be pantheistic. Ours could be a world whose events were somehow
forced to possess causal orderliness without its being true that hammers *were intrinsically such that* they could move nails when they hit
them, which is what various philosophers have held. It could instead
be that the world's orderly patterns were intrinsically worth contemplating by a divine mind and that this was what explained both the
existence of hammers and their orderly behaviour. Why would the
divine mind itself exist? Chapter 5 tried to show how a modern
Platonic answer might run.

'But don't we have immediate insight into how a hammer can drive
a nail? Aren't hammers hard, and massive compared with nails?' This
is just our knowing from experience how the world runs. It's like
knowing that hammers fall when you let go of them. ('They're heavy,
aren't they?') A child feels sure of understanding hammers but finds
magnets magical. A physicist, in contrast, recognizes that being hard
and being massive are affairs complex enough to puzzle the best
brains, and that a hammer's earthwards movement contains as much
mystery as a nail's movement towards a magnet. The most fundamental laws known to us humans have as little intrinsic obviousness as any
law that incantations can make broomsticks fly or that devils can cause
fires by merely wanting them to break out.

The Platonic approach, I argued, could give us some real understanding of why things of certain kinds are more than merely possible.
Things can be intrinsically good, meaning that their existence can satisfy requirements of an absolute kind. Maybe such requirements
would never be able to carry responsibility for their own fulfilment, not
even in the case of divine minds of infinite goodness, but if so then this
would be an affair not of logically demonstrable necessity but of synthetic necessity (see Chapter 5); now, it would be equally simple if the
facts of synthetic necessity were instead such that at least one ethical
requirement did carry responsibility for its own fulfilment, making
Plato right. Why various things were more than merely possible could
in that case have an explanation which wasn't totally beyond our

grasp. But what real understanding could we have *of one thing's being able to bring about another* if Plato's approach couldn't be made relevant in, say, the fashion in which pantheists could make it relevant?

Here, let's imagine, we have *an actual thing*: a person willing something, or a magnet, or a moving hammer. There is in addition *a merely possible thing or event*: a universe such as the person is trying to will into existence, or the event of two dice falling double-six which the person is willing to occur, or the movement of a nail towards the magnet or ahead of the hammer. Now, what equips the actual thing with productive power over the mere possibility? It can seem easy enough to understand how things *once they do exist* can stand in necessary relationships. Two rods are placed side by side. If they have the same length then, just through each being the rod that it is, they stand in the relationship of being-of-the-same-length *necessarily*. A rod-creating deity wouldn't have to create the rods *and next* the sameness of their lengths. Yet how about the relationship of having-power-to-produce in which one thing or event is supposed to stand to some other thing or event, this other thing or event being one *that isn't there yet* (for otherwise there'd be no need to produce it)? How on earth could this relationship exist necessarily? You can see why defenders of the Regularity Theory of Causation thought there was no alternative to their position.

Cosmic Fine Tuning

Intelligent life is a product of billions of years of Darwinian evolution. How comes it that our cosmic environment remained favourable enough during all those years? Why, in fact, were its properties such that intelligent life could ever have evolved in it? Had the expansion speed early in the Big Bang been slower by one part in a billion then, it is often held, our universe would almost immediately have collapsed under the influence of gravity. With an equally trivial speed increase there would soon have been nothing but very cold, very dilute gases, unable to form any life-giving stars. Again, the early Bang was immensely smooth rather than turbulent: smooth to (as calculated by Roger Penrose) one part in 1 followed by a thousand trillion trillion

trillion trillion trillion trillion trillion trillion trillion trillion zeros, a number whose size you begin to appreciate when you reflect that the number of atoms in the volume scanned by our telescopes is only about 1 followed by eighty zeros. Were it not for its early smoothness, our universe would quickly have come to consist just of black holes; or, had this outcome somehow been avoided, then everything would have remained immensely hot for billions of years, after which all would have been so far spread out that once again no stars could have formed. Now, some unknown law of physics may perhaps have dictated the smoothness; but couldn't we instead see it as a sign of divine action? Many scientists have concluded that ours is a universe remarkably *fine tuned for life*, meaning that it is characterized by sets of figures—the relative strengths of its physical forces, the relative masses of its elementary particles, its early expansion speed, and so on—such as could easily have been different and such that tiny alterations in them would have prevented the evolution of living beings of any kind. And although talk of 'being fine tuned for life' means simply what I have just now said it meant instead of being tied to the idea of a divine Fine Tuner, various of the scientists do see those sets of figures as evidence of God's reality.

Claims about the fine tuning are often very controversial. Thus, the early cosmic expansion speed and degree of smoothness might perhaps have been made inevitable by a process known as 'inflation'. Perhaps cosmologists are wrong when they describe the Big Bang as never having done anything but slow down. Perhaps there was, very early on, a sudden and extremely brief burst of *accelerating* expansion during which everything became many trillion times more spread out. Any early roughness might then have become smoothed away, much as the wrinkles on a balloon's surface disappear when the balloon is inflated, and the subsequent expansion speed could automatically have become just right for life's purposes. Yet some scientists argue that the conditions necessary for inflation (or at least for inflation of a propitious kind, able to yield density fluctuations suitable for seeding the growth of galaxies) were themselves in need of extremely accurate tuning, while others such as Penrose think no plausible amount of inflation would have sufficed.

To my mind what's impressive is not any particular argument for viewing this or that as an instance of fine tuning. Instead it is that such arguments exist in such numbers, plus the availability of two fairly straightforward means of explaining any fine tuning, ways each involving the notion of *a selection effect*. As will be examined in a moment, it could be that something other than *divine selection* of universe-properties was involved. *Observational selection* would be an alternative because—this is the Anthropic Principle as Brandon Carter (its originator) understands it—all living beings must find themselves in life-permitting universes, mustn't they? But before discussing the point let me list various factors which were remarkably well tuned for life's purposes, according to claims made by many people. (To defend these and other claims would be a very lengthy task. I attempted it in *Universes* (1989*a*) with numerous references to the scientific literature.[3])

1. As mentioned above, one could point to the apparent tuning of the cosmic expansion speed and degree of smoothness, or alternatively to the tuning of the conditions needed for inflation to occur at all, or for it to occur propitiously. 'Bare lambda' and 'quantum lambda', two components of an inflation-driving 'cosmological constant', would have needed to come to cancel each other with immense precision. Some have argued that a minuscule divergence (perhaps of far less than one part in a trillion) from the actual strength ratio in which gravity stands to the nuclear weak force would have prevented the cancellation, and that this would have meant that inflation couldn't have ended. Again, the masses of numerous 'scalar particles' may have needed to be tuned very accurately for inflation to take place and then terminate appropriately.

2. Carbon, crucial (for reasons people think they can understand) to all known life forms, is manufactured abundantly by stars, so

[3] Apart from Leslie 1989*a*, see several articles in Leslie 1990*a*, particularly those by Bernard Carr and Brandon Carter; or perhaps Leslie 1982, 1988*b*, or 1994*d* and *e*. Also Atkins 1981; Balashov 1991; Barrow and Tipler 1986; Carr and Rees 1979; Carter, in Bertola and Curi 1993: 33–66; Davies 1982; Demaret and Barbier 1981; Ellis 1993; Polkinghorne 1986; Rees 1997, 1999; Rozental 1980, 1988; Smolin 1997: 300–15 and 324–6.

that it can then be distributed widely through stellar explosions. Well, why? It is thanks to the carbon nucleus just managing to 'resonate' appropriately, while the oxygen nucleus just fails to resonate in a carbon-destroying manner. With a strength increase of perhaps 1 per cent in the nuclear strong force, almost all of a star's carbon would be converted to oxygen. Again, tiny alterations in this same force, or in Planck's constant, or in the mass difference between the neutron and the proton, would seemingly have led to a universe almost entirely without chemistry, or would have made stars burn either trillions of times faster or not at all, or would have turned even small bodies into miniature neutron stars.

3. Fairly small alterations in the strength of the nuclear weak force would have destroyed the hydrogen that was needed for making steadily burning stars and water, or would have prevented the proton–proton and carbon–nitrogen–oxygen cycles that make stars into sources of elements heavier than helium. Again, this force had to be neither too strong for neutrinos to escape from a supernova's collapsing core, nor too weak for them to blast off its outer layers, so scattering into interstellar space (for later formation into planets) atoms such as you and I are made of.

4. Slight strengthening of electromagnetism would seemingly have made atoms impossible (by transforming all quarks into leptons) or would have led to rapid decay of all protons (even quite a slow decay rate would render our bodies violently radioactive) or would have caused protons to repel one another so strongly that hydrogen became the only possible element, or would have ensured that chemical changes were immensely slow.

5. The relative strengths of gravity and electromagnetism may have needed tuning to one part in many trillion for there to be stars like the sun: stars burning steadily for billions of years and capable of providing warmth and light of suitable wavelengths to planets positioned at safe distances. Think, here, of how extremely destructive short wavelengths can be, and of how, in contrast, the red light in a photographer's darkroom fails to

affect photographic film because its individual photons pack too little punch for this—or for the photosynthesis on which plants depend. Bear in mind, too, that in order to be to be life-bearing a planet must presumably be far enough from its star to be largely unaffected by stellar flares, also avoiding gravitational locking such as makes our moon turn always the same face towards us.

6. Various superheavy particles were important early in the Bang. Had their masses not fallen inside fairly narrow limits, then the universe would soon have been composed of light rays and hardly any matter, or else there would have been so much matter that instead of stars and planets there would have been black holes only.

7. Had the mass of the 'top' quark been only a little greater, then this would have led to what is known as 'a vacuum instability or scalar field disaster'.[4]

Philosophical Objections to Arguments Based on Fine Tuning

Attempts to present the alleged fine tuning as a reality, or as genuine evidence of anything interesting, are often opposed on grounds abstract enough to count as 'philosophical'. Let us very briefly examine some of these. (*Universes* considered them and others in more detail.[5])

(*a*) 'Why', it is protested, 'should anybody view *life*, or *intelligent life*, as specially in need of explanation? Why not talk instead of fine tuning for carbon, for instance? Carbon is in its way remarkable through being able to form both diamonds and graphite, but who would dream of building any Proof of God on *that*?' To this the best reply could be an appeal to the Merchant's Thumb Principle: the principle that a good reason for suspecting that something calls for explanation is that

---

[4] To get some understanding of these last mysterious words, see Ellis *et al.* 1990 (already referred to in Ch. 4).

[5] Or see Leslie 1983*a* or perhaps 1983*c*, 1985*a*, 1986*a* and *b*, 1988*a* and *c*, or 1997*e*. Also Richard Swinburne, 'Argument from the Fine-Tuning of the Universe', in Leslie 1990*a*.

an attractive explanation comes to mind. (Why does the thumb's position specially need to be explained? In presenting a silk robe for a customer's scrutiny, don't one's thumbs have to be *somewhere*? Well, this particular thumb is so placed that it conceals a hole in the silk.) Suppose you have caught a fish measuring 17.82 inches. 'What's so remarkable in that? Every fish must have some length or other!' Yes, but what if you find your fishing apparatus to be so constructed that it could catch none but fish of this length, plus or minus very little? You could then rightly be inclined to seek some special explanation, such as that somebody powerful and benevolent wanted you to have a fish supper. The point of all this is that we could well think we saw why a divine creative force or person would be particularly concerned with creating the kind of universe in which intelligent life could evolve. This could suggest an attractive way of accounting for any fine tuning— provided, that's to say, that 'the God hypothesis' didn't have, prior to our attention being drawn to the fine tuning, too low a probability of being right.

(b) It is suggested that people impressed by so-called evidence of fine tuning are being sadly unimaginative. Isn't it easy enough to picture life-forms for which no tuning would be needed? Isn't awe at how things are fine tuned *for life as we know it* like gasping at how the Mississippi manages to thread its way under every single bridge? May it not be like the wonder felt by arsenic-eating beings at how their planets are so plentifully supplied with arsenic? My response to this is that, yes, we cannot entirely rule out life able to exist on neutron stars through being based not on chemistry (which means on electromagnetism) but on the nuclear strong force; or plasma life deep inside the sun; or life consisting of crystalline patterns in frozen hydrogen; or life taking the form of intricately organized interstellar clouds; or the various other strange types of life imagined in *Life Beyond Earth,* a fascinating book by Gerald Feinberg and Robert Shapiro (1980). Yet, first, believing in such life (and above all in its being complex enough to be intelligent) could be less reasonable than belief in divine selection or observational selection; and second, as has been insisted by I. L. Rozental, one would appear to need a great deal of fine tuning in order to get a universe of neutron stars, suns, frozen hydrogen, interstellar clouds. A universe collapsing almost at once, or one that quickly came to consist only of extremely cold and dilute gases, or a universe composed almost entirely of black holes or of light rays, could be much more to be expected.[6]

(c) 'Much more to be expected'? Had our universe *not* been suited to the evolution of intelligent life then, it is sometimes protested, there'd be nobody to discuss the affair—which, the protestor argues, means there could be nothing remarkable here. Yet, I ask, suppose you had faced a firing squad of fifty sharpshooters, all of whom had missed you; wouldn't you have reason for believing you were popular with those sharpshooters? Would you dream of saying to yourself, 'Their failure to hit me, *since it was necessary to my now being able to think anything*, of course cannot supply me with grounds for any conclusion'?

(d) People sometimes object that in this area there couldn't be 'the repetitions that are needed before anything can be judged *probable or improbable*'; for, they say, the universe *is something unique*. But why couldn't there be repetitions, I ask? Aren't multiple-universe scenarios common in today's journals of physics and cosmology? Again, don't articles in these journals often make the point that many very important properties of our universe could have resulted from how, very early in the Big Bang, various quantum events just happened to turn out, *the probabilities* of their turning out in the one way or in the other having been laid down by the laws of quantum theory? And if the line of reasoning proved anything, wouldn't it prove that there would be nothing remarkable in a world whose every cliff bore the words 'God exists' or, for that matter, the entire Bible or Koran? Perhaps the people running such reasoning have forgotten that a thing that was unique under one description (such as the description 'universe') could still be non-unique under other descriptions (such as 'thing governed by the probabilistic laws of quantum theory' or 'thing with properties suggestive of intelligent design').

(e) Rather more interestingly, it is argued that even when the word 'universe' is defined (as is nowadays typical) in a way making it

---

[6] Feinberg and Shapiro 1980, or see the article by the same authors in Leslie 1990a; Rozental 1980.

conceivable that there actually exist many universes, still *only a single universe is open to our inspection*. How, then, could we possibly judge whether its properties stood in special need of explanation? Perhaps talk of 'fine tuning' is totally inappropriate because some as-yet-undiscovered fundamental theory dictated every force strength, every particle mass, and everything else. But while this point may have some force, it seems to me far from sufficient to ruin all evidence of fine tuning. Suppose we had found that electromagnetism was stronger than gravity by a factor of 112,012,100,100,202,100, 021,011,211,021,112,100, a figure which spells out MADE BY GOD when its zeros, ones, and twos are interpreted as Morse-code dots, dashes, and spaces. What would we think of the argument that, being able to see only a single universe, we couldn't justifiably conclude that this suggested anything? (Entering a room, you see a computer screen with numbers appearing on it. You are told the computer is calculating the value of *pi*. The first digits you see are the kind of apparent rubbish you would expect: perhaps 915484612235006. Then you see the figure mentioned just now, easily interpreted as the message 'Made by God'. How would you react to the suggestion that this figure had been in no way 'improbable' or 'fine tunable' because all its digits were instead mathematically necessary elements of *pi*, the eternally fixed ratio of a circle's circumference to its diameter? Surely a mischievous computer programmer would be far easier to believe in.)

(f) Sometimes it is objected that we cannot examine the entire range of logically possible laws of nature and values of physical constants (ratios between force strengths, etc.) so as to determine what proportion of the possibilities would correspond to life-permitting universes. But my answer to this is that no such task need be undertaken by believers in fine tuning. All they need consider is the possible universes 'in the local area': ones obeying basic laws recognizably like those with which we are acquainted, and differing from our universe not too greatly in the relative strengths of their physical forces, the relative masses of their elementary particles, and so forth. (A fly sits on a wall. In a fairly large area around the fly—perhaps it stretches a yard in all directions, but what's crucial is simply that it is large by comparison with the fly itself—we see no flies or other objects bigger than dust grains. Well now, suppose a bullet hits the fly, and that no others have hit the fairly large area in question. A natural thought would be that very probably some marksman aimed the bullet. One could think this without forming any theories about whether distant areas of the wall, or other quite different walls, were so covered with flies that almost any bullet hitting *there* would hit one. What's important is that *the local area* contains only a single fly. The moral of this tale is that we can reasonably be impressed by how *comparatively minor* alterations to such things as force strengths would seemingly have meant that no life would ever have evolved in our universe. Whether universes would be life-permitting if their force strengths, etc., were immensely different, or if they were governed by very different fundamental laws, need be no concern of ours.)

Fine Tuning might Suggest Observational Selection

As mentioned earlier, any 'fine tuning for life' found in our universe's properties might be a sign of something other than divine selection. *Observational selection* could be involved instead. Physicists and cosmologists have proposed numerous mechanisms for generating regions of reality which—in part because these various regions would be largely or entirely separated from one another, and in part because their characteristics are pictured as varying widely in crucial respects—could be worth calling 'many distinct universes'. While their most fundamental laws could be always exactly the same, their force strengths, particle masses, expansion speeds, and so forth, could differ for reasons that can readily be guessed. Given sufficiently many, sufficiently diversified universes it could be more or less inevitable that at least a few universes would be tuned appropriately for life's purposes. We living beings could then find ourselves only in a universe among those perhaps very rare ones.

I examined possible universe-generating and universe-diversifying mechanisms in detail in *Universes*, with references to the writings of many scientists.[7] Inside a cosmos whose extent could be anything up to infinite, universes might be domains so huge that an observer deep inside any one of them couldn't see the others. They could differ markedly in their expansion speeds, in their degrees of turbulence, and in many other features. Alternatively, universes might exist as successive cycles of an oscillating cosmos; or they might be regions which had come to form the separate worlds of Hugh Everett's many-worlds quantum theory, regions jostling one another in ways detectable by physicists (in the double-slit experiment, for example). Again, they could be new zones of comparatively calm expansion to which a violently inflating cosmos was continually giving birth; or they could be buds pinched off from other universes; or they might spring into existence entirely independently as was first suggested by Tryon. Once you have a plausible universe-generating mechanism, it can be strange to think it has operated *only once*. To stop it operating again and again, what complex factors you could have to dream up!

Why might the properties of the various universes differ? Several possible answers have been worked out by physicists. A particularly attractive one runs as follows. Early in all of the perhaps infinitely many Big Bangs that have occurred, each the birth of a new universe, conditions would have been so hot that there was (at least for all practical purposes) only a single force and a single type of elementary particle. Very shortly afterwards this simple situation was destroyed through the action of one or more *scalar fields* which appeared as things began to cool. The intensity of any such field might vary randomly from place to place for a reason as straightforward as this: that markedly different intensities could all possess more or less the same potential energy—potential energies being what physical systems want to minimize in the sense in which a ball wants to be at bottom of a valley. Now, the scalar field or fields could be very important through interacting with elementary particles, giving them mass where before they were

---

[7] In addition, see the articles by George Gale, Andrei Linde, Edward Tryon, and John Wheeler in Leslie 1990*a*. Also Barrow and Tipler 1986; Davies 1982; Halliwell 1992; Rees 1997; Rozental 1988.

massless (as in the case of photons, Newton's 'particles of light', which are massless even nowadays until they encounter the fields inside superconductors). And as the masses of various particles varied from one particle type to another in any one place, or from one place to another in any large situation split into domains whose scalar field intensities were different, so also the effective strengths of physical forces would vary. (If a particle is heavier, a force finds it harder to push around; and again, the effective ranges of various forces depend on the masses of the 'messenger particles' that convey them.) What is more, early cosmic inflation could have meant that any scalar field, when it had once possessed the same intensity throughout even an extremely tiny region, would come to have an intensity that was the same everywhere inside in a domain much greater than the one visible to our telescopes. Domains of this kind, each characterized by a different mix of particle masses and force strengths, could well deserve to be called 'universes'.

People might therefore think that observational selection was fully equal to divine selection in its ability to throw light on any fine tuning. Wouldn't it then follow that whatever evidence we had of fine tuning would be of no use for supporting belief in God? Not at all. The evidence could support—that is, increase the probable correctness of— *both* the God hypothesis *and* the hypothesis that there exist many and varied universes. If Andy is robbed on an island whose other inhabitants are Billy and Charlie, then this increases the probability that Billy is a thief, and does exactly the same thing to Charlie.

However, there could be various factors tending to point towards divine selection rather than observational selection. The next section will look at this possibility.

Matters Hard to Treat as Observationally Selected

As a means of throwing light on our fascinatingly structured world, how does the God hypothesis fare by comparison with the observational selection hypothesis? (1) For a start, let us not forget that the evidence of fine tuning doesn't stand alone. The God hypothesis might throw light on much more than the fine tuning. It might explain why there exists any world at all, and why there is anything worth calling a system of causal laws. (2) Next, we can draw attention to several seemingly fortunate affairs that are *less readily viewable as products of randomization* than any particle mass or force strength or speed of expansion.

Here the crucial point is that randomization of such things as particle masses wouldn't necessarily involve variation in Nature's fundamental laws. It might instead be, as many people think, that scalar fields which differed from place to place compelled particle masses to vary as well, and assorted other mechanisms could have been equally good candidates for producing such differences against a background of laws which never varied. For instance, some argue that interactions between universes in a tremendously complicated 'foamy' situation could randomize the values of all manner of physically important parameters inside the various universes. While all this is very speculative, it can claim some scientific respectability. In contrast, the theory that fundamental laws themselves vary from universe to universe would seem one for which no scientific arguments could be given. Scientists put considerable trust in the principle of induction: the principle that what has been observed so far should be taken as typical, more or less, of what takes place elsewhere. Sure enough, new circumstances can result in markedly novel events against a background of unvarying laws, as good scientists know. Pour radioactive liquid from a cylindrical container into a spherical one and you may suddenly get an explosion. Still, how about the idea that Nature's most basic laws differ from one location to another? This can look a very different matter. Perhaps it wouldn't be in utter conflict with the scientific way of thinking, so that a theistically minded scientist might accept it, provided that what were being pictured were divinely produced differences between regions separate enough to be counted as 'separate universes'. None the less it would be something that science itself could not support.

Seeing, however, that so hugely many other fundamental laws could be described without self-contradiction, might it not be terribly parochial to fancy that the ones ruling our universe ruled all further universes as well? I think not. It seems to me the very essence of inductive reasoning that we should favour such 'parochialism'.

'Still', you might protest, 'if all possible laws reigned somewhere or other, wouldn't this be a very satisfactorily simple situation in one respect? Just look at how few words—*All possible laws reign somewhere*—it would take to identify it!' Now, that's a point in which I see at least some slight force. But there are also the very forceful points (i) that such a situation would be in another way extremely complex, obviously, and (ii) that a liking for its special brand of simplicity could be the first step towards thinking that not just all possible law-controlled situations, but all possible situations whatsoever, were in existence somewhere—which would be the death of all inductive reasoning if the arguments of Chapter 1 were right.

If, then, there were aspects of Nature's workings that appeared very fortunate and also entirely fundamental, then these might well be seen as evidence specially favouring belief in God. (Combining the idea of observational selection with that of *variation of fundamental laws* from universe to universe *would not* be a fine means of generating 'a scientific alternative to theism', I think.) What might such aspects be? Well, we might point to the principle of special relativity. Regardless of whether a force such as electromagnetism is acting at right angles to a system's direction of travel, its effect is invariable thanks to this principle; the force tugs as strongly east–west as north–south. The principle can be thought rather a strange one. Who would have dreamt that light could overtake us at the same measured speed regardless of whether we had accelerated violently throughout last week? Yet without such a principle to keep forces acting invariably, genetic codes could be unworkable and planets could too easily break up as they rotated.

Look, too, at the laws of quantum theory. These stop electrons spiralling into atomic nuclei. They permit apparently dissipated wave-energy to be released in concentrated bursts so that it can work usefully, as in the case of photosynthesis. They explain why particles do not wander all over the place, as was shown by Richard Feynman. They are also behind the fact that atoms come in standardized types, something crucial not only to genetic codes but to the functioning of all of Earth's organisms. And like the laws of relativity they can well seem completely basic among the laws ruling our universe.[8]

Believers in God could also point to the curious fact that a physical force strength or an elementary particle mass often appears to require tuning to a given figure, plus or minus very little, *for several different reasons at once*. Look once again at electromagnetism. Electromagnetism could seem to have required tuning, sometimes with extreme accuracy, (i) for there to be stars like the sun, steadily burning for billions of years and giving off warmth and light of the right kinds (no grilling of organisms by excessive amounts of ultraviolet light, for example); (ii) for stellar carbon synthesis and for the processes which produce supernovae that scatter carbon and other crucial elements through the galaxy; (iii) for quarks not to be replaced by leptons, making atoms impossible; (iv) for protons not to decay swiftly; (v) for protons not to repel one another so strongly that there could be no chemistry; (vi) for chemical changes to occur at appropriate rates (a strength increase by as little as 1 per cent could have doubled the number of years needed for intelligent life to evolve); and for various other reasons as well. (For example, John Barrow and Frank Tipler argue on quantum-theoretical grounds that the smallness of the electromagnetic fine structure constant is essential to there being any fairly firm distinction between matter, from which you can make organisms, and radiation, from which you can't.) Now, how come that *the same one force strength* manages to satisfy so many different requirements? Why doesn't electromagnetism need tuning to one strength for there to be stars that burn appropriately, to another for carbon synthesis, to a third for chemistry to be possible, and so on? What was needed here, it could be tempting to think, was divine ingenuity in selecting fundamental laws of physics that did not run into this problem, or the kind of substitute for divine ingenuity in which Plotinus believed.

---

[8] Ch. 3 of *Universes* (Leslie 1989*a*) took this line of thought further by discussing the principle of baryon conservation ('mediated by no force field, this none the less manages to prevent the universe becoming a fireball of radiation'), the mystery of particle spins (without which 'there would be neither electromagnetism nor gravity', as was argued by Rozental), and the perhaps still greater mystery of why the world has 'an arrow of time' (I quoted Penrose's remark that we are here 'groping at matters that are hardly understood at all from the point of view of physics').

Observational Selection Inside a Pantheistic Scheme of Things

Might it not be, though, that belief in God could actually be combined with belief in multiple universes and observational selection? It is sometimes suggested that God would create not only innumerable life-containing universes, thereby benevolently maximizing the number of living beings, but also countless others *that weren't* life-containing, for the sheer pleasure of watching them evolve. Could this make sense?

I suggest not, *if one assumes that the universes would be outside God,* and if God is to be pictured as knowing everything worth knowing. What sense can there be in the idea of a deity who knows everything worth knowing but who then has to create this or that universe so as to be able to see its beauty and grandeur? Why not just view it 'with his mind's eye'? We humans can do better when we actually produce things, because the results of our attempts to imagine them are so defective, but God is supposed not to share our limitations. God is meant to be able to grasp exactly what universes developing in accordance with various laws would be like, without actually creating anything outside himself and then looking at it. In the case of indeterministic laws, God is meant to be able to picture all possible results of their operation.

However, this book has argued that it could be far better to conceive our universe and innumerable others as existing *inside* a divine mind. No scientific findings could refute this, for what is being suggested is that scientific experiments and scientists and all the other things in our universe are parts of the divine reality. The divine reality is consciousness of everything worth knowing, and the structure of our universe is well worth knowing. Its being contemplated by a divine mind in all its details makes it the structure not just of something possible, but of something actually existent. Now, *perhaps not every universe worth thinking about is a life-containing universe.* Any divine mind could be the richer for considering various universes that were lifeless.

The situation could be summarized as follows.

A: In the absence of God, the combination of multiple universes with observational selection—selection set up by the fact that living

215

beings must always find themselves in life-permitting universes—probably couldn't make it unmysterious that our universe is so well suited to life's evolution. This is because of the various matters which seem unachievable by randomization of a kind for which a scientist could argue when reasoning simply as a scientist. They concern the suitability of fundamental laws, and science could never itself supply grounds for thinking that *those* varied randomly from place to place.

B: According to pantheism, though, divine thinking could well extend to the structures of hugely many universes which weren't life-containing—and which, for that matter, might in many cases be governed by fundamental laws that were firmly life-excluding. All these universes would then (says pantheism) actually exist. Why would we living beings find ourselves in a life-permitting universe? Answer: because of observational selection.

Bibliography

Alston, W. P. (1986). 'Does God have Beliefs?' *Religious Studies* (Sept.): 287–306.

Armour, L. (1992). *Being and Idea: Developments of Some Themes in Spinoza and Hegel* (Hildesheim and New York: Georg Olms Verlag).

Atkins, P. W. (1981). *The Creation* (Oxford: W. H. Freeman).

Ayer, A. J. (1970). *Metaphysics and Common Sense* (San Francisco: Freeman, Cooper).

Balashov, Y. V. (1991). 'Resource Letter AP–1: The Anthropic Principle', *American Journal of Physics* (Dec.): 1069–76.

Barrow, J. D., and Tipler, F. J. (1986). *The Anthropic Cosmological Principle* (Oxford: Clarendon Press).

Bergson, H. (1935). *The Two Sources of Morality and Religion*, tr. R. A. Audra and C. Brereton (Garden City, NY: Doubleday).

Bertola, F., and Curi, U. (eds.) (1993). *The Anthropic Principle* (Cambridge: Cambridge University Press).

Block, N. (1978). 'Troubles with Functionalism', in C. W. Savage (ed.), *Perception and Cognition: Minnesota Studies in the Philosophy of Science*, ix (Minneapolis: University of Minnesota Press), 261–325.

Bohm, D. (1990). 'A New Theory of the Relationship of Mind and Matter', *Philosophical Psychology*, 3/2: 271–86.

—— and Hiley, B. J. (1993). *The Undivided Universe* (London and New York: Routledge).

Bradley, F. H. (1893). *Appearance and Reality* (London: Oxford University Press).

—— (1914) *Essays on Truth and Reality* (Oxford: Oxford University Press).

—— (1927). *Ethical Studies*, 2nd edn., revised (Oxford: Clarendon Press).

—— (1935). *Collected Essays*, ii (Oxford: Oxford University Press).

Broad, C. D. (1925). *The Mind and its Place in Nature* (London: Routledge & Kegan Paul).

Cantor, G. (1932). *Gesammelte Abhandlungen*, ed. A. Fraenkel and E. Zermelo (Berlin: Springer-Verlag).

Carr, B. J., and Rees, M. J. (1979). 'The Anthropic Principle and the Structure of the Physical World', *Nature* (12 Apr.): 605–12.

Bibliography

Clark, S. R. L. (1990). 'Limited Explanations', in D. Knowles (ed.), *Explanation and its Limits* (Cambridge: Cambridge University Press), 195–200.

Clifford, W. K. (1874). 'Body and Mind', in his *Lectures and Essays*, ed. L. Stephen and F. Pollock (London: Macmillan, 1886), 244–73.

Craig, W. L. (1999). 'Timelessness, Creation, and God's Real Relation to the World', *Laval théologique et philosophique* (Feb.): 93–112.

Crandall, R. E. (1997). 'The Challenge of Large Numbers', *Scientific American* (Feb.): 74–8.

Curley, E. (ed. and tr.) (1985). *The Collected Works of Spinoza*, i (Princeton: Princeton University Press).

Dauben, J. W. (1979). *Georg Cantor: His Mathematics and Philosophy of the Infinite* (Boston: Harvard University Press).

Davies, B. (1997). 'Aquinas, God and Being', *The Monist* (Oct.): 500–17.

Davies, P. C. W. (1980). *Other Worlds* (London: Dent).

—— (1982) *The Accidental Universe* (Cambridge: Cambridge University Press).

Demaret, J., and Barbier, C. (1981). 'Le Principe anthropique en cosmologie', *Revue des questions scientifiques* (Oct.): 461–509.

Dennett, D. C. (1991). *Consciousness Explained* (Boston: Little, Brown).

Deutsch, D. (1985a). 'Quantum Theory, the Church-Turing Principle and the Universal Quantum Computer', *Proceedings of the Royal Society, London* (July): 97–117.

—— (1985b). 'Quantum Theory as a Universal Physical Theory', *International Journal of Theoretical Physics* (Jan.): 1–41.

—— (1997). *The Fabric of Reality* (London: Allen Lane).

Eddington, A. S. (1928). *The Nature of the Physical World* (Cambridge: Cambridge University Press).

Edwards, D. L. (ed.) (1963). *The Honest to God Debate* (London: SCM Press).

Edwards, P. (ed.) (1967). *The Encyclopedia of Philosophy* (New York: Macmillan and Free Press).

—— (ed.) (1992). *Immortality* (New York: Macmillan).

—— (1996). *Reincarnation: A Critical Examination* (Amherst, NY: Prometheus Books).

Einstein, A. (1962). *Relativity: The Special and the General Theory* (15th edn., enlarged; London: Methuen).

Ellis, G. F. R. (1993). *Before the Beginning* (London: Bowerdean Press).

Ellis, J., Linde, A., and Sher, M. (1990). 'Vacuum Stability, Wormholes, Cosmic Rays and the Cosmological Bounds on m_t and m_h', *Physics Letters B* (13 Dec.): 203–11.

218

Bibliography

Everett, H. (1957). '"Relative State" Formulation of Quantum Mechanics', *Reviews of Modern Physics* (July), 454–62.

Ewing, A. C. (1973). *Value and Reality* (London: George Allen & Unwin).

Feinberg, G., and Shapiro, R. (1980). *Life Beyond Earth* (New York: William Morrow).

Forrest, P. (1993). 'Difficulties with Physicalism, and a Programme for Dualists', in H. Robinson (ed.), *Objections to Physicalism* (Oxford: Clarendon Press), 251–9.

—— (1996). *God without the Supernatural* (Ithaca, NY, and London: Cornell University Press).

Foster, J. (1982). *The Case for Idealism* (London: Routledge & Kegan Paul).

Fröhlich, H. (1968). 'Long-Range Coherence and Energy Storage in Biological Systems', *International Journal of Quantum Chemistry*, 2/5: 641–9.

—— (1986). 'Coherent Excitation in Active Biological Systems', in F. Gutman and H. Keyzer (eds.), *Modern Bioelectrochemistry* (New York: Plenum), 241–61.

—— and Kremer, F. (eds.) (1983). *Coherent Excitations in Biological Systems* (Berlin: Springer-Verlag).

Gale, R. M. (ed.) (1967). *The Philosophy of Time* (Garden City, NY: Doubleday).

—— and Pruss, A. R. (eds.) (2002). *The Existence of God* (Aldershot: Ashgate).

Garrett, D. (ed.) (1996). *The Cambridge Companion to Spinoza* (Cambridge: Cambridge University Press).

Gershenfeld, N., and Chuang, I. L. (1998). 'Quantum Computing with Molecules', *Scientific American* (June): 66–71.

Grim, P. (1991). *The Incomplete Universe* (Cambridge, Mass.: The MIT Press).

Grover, S. (1998). 'Cosmological Fecundity', *Inquiry* (Sept.): 277–99.

Grünbaum, A. (1960). 'Logical and Philosophical Foundations of the Special Theory of Relativity', in A. Danto and S. Morgenbesser (eds.), *Philosophy of Science* (Cleveland: World Publishing Co.), 399–434.

—— (1967). 'The Status of Temporal Becoming', in Gale (1967: 322–53).

—— (1973). *Philosophical Problems of Space and Time* (2nd edn., enlarged; Dordrecht and Boston: Reidel).

—— (2000). 'A New Critique of Theological Interpretations of Physical Cosmology', *British Journal for the Philosophy of Science* (Mar.): 1–43.

Gutman, J. (ed. and tr.) (1949). *Benedict de Spinoza: Ethics and On the Improvement of the Understanding* (New York: Hafner Press).

Halliwell, J. J. (1992). *Quantum Cosmology* (Cambridge: Cambridge University Press).

Bibliography

Hameroff, S. R. (1974). 'Chi: A Neural Hologram?', *American Journal of Clinical Medicine*, 2/2: 163–70.

—— (1994). 'Quantum Coherence in Microtubules: A Neural Basis for Emergent Consciousness?', *Journal of Consciousness Studies*, 1/1: 91–118.

Hartle, J. B., and Hawking, S. W. (1983). 'Wave Function of the Universe', *Physical Review D* (15 Dec.), 2960–75.

Hartshorne, C. (1948). *The Divine Relativity* (New Haven: Yale University Press).

—— (1970). *Creative Synthesis and Philosophic Method* (London: SCM Press).

—— (1984). *Omnipotence and Other Theological Mistakes* (Albany, NY: State University of New York Press).

Hawking, S. W. (1988). *A Brief History of Time* (New York: Bantam).

—— (1993). *Black Holes and Baby Universes* (New York: Bantam).

Heidmann, J. (1992). *Intelligences extra-terrestres* (Paris: Odile Jacob).

Hepburn, R. W. (1988). 'The Philosophy of Religion', in *An Encyclopaedia of Philosophy* (London: Routledge), 857–77.

Hiley, B. J., and Peat, F. D. (1987). *Quantum Implications* (London and New York: Routledge).

Hodgson, D. (1991). *The Mind Matters* (Oxford: Clarendon Press).

Hume, D. (1739). *A Treatise of Human Nature* (London; repr. 1888, ed. L. A. Selby-Bigge, London: Oxford University Press).

James, W. (1890). *The Principles of Psychology*, i (New York: Henry Holt; repr. 1950, New York: Dover).

—— (1912). *Essays in Radical Empiricism* (New York: Macmillan).

Jantzen, G. M. (1984). *God's World, God's Body* (London: Darton, Longman & Todd).

Jeans, J. (1930). *The Mysterious Universe* (London: Macmillan).

Kane, R. H. (1976). 'Nature, Plenitude and Sufficient Reason', *American Philosophical Quarterly* (Jan.): 23–31.

King-Farlow, J. (1978). *Self-Knowledge and Social Relations* (New York: Science History Publications).

Küng, H. (1980). *Does God Exist?* (London: Collins).

Leclerc, I. (1981). 'The Metaphysics of the Good'. *Review of Metaphysics* (Sept.): 3–25.

—— (1984). 'God and the Issue of Being', *Religious Studies* (Mar.): 63–78.

Leibniz, G. W. (1714). *New Essays on Human Understanding*, tr. J. Bennett and P. Remnant, 1981, Cambridge: Cambridge University Press.

Leslie, J. (1970). 'The Theory that the World Exists Because it Should', *American Philosophical Quarterly* (Oct.): 286–98.

—— (1971). 'Morality in a World Guaranteed Best Possible', *Studia Leibnitiana*, 3/3: 199–205.

—— (1972). 'Ethically Required Existence', *American Philosophical Quarterly* (July): 215–24.

—— (1973). 'Does Causal Regularity Defy Chance?', *Idealistic Studies* (Sept.): 277–84.

—— (1976*a*). 'The Value of Time', *American Philosophical Quarterly* (Apr.): 109–21.

—— (1976*b*). 'The Best World Possible', in J. King-Farlow (ed.), *The Challenge of Religion Today* (New York: Neale Watson), 43–72.

—— (1978*a*). 'God and Scientific Verifiability', *Philosophy* (Jan.): 71–9.

—— (1978*b*). 'Efforts to Explain All Existence', *Mind* (Apr.): 181–94.

—— (1979). *Value and Existence*. Oxford: Blackwell.

—— (1980). 'The World's Necessary Existence', *International Journal for Philosophy of Religion* (Winter): 207–23.

—— (1982). 'Anthropic Principle, World Ensemble, Design', *American Philosophical Quarterly* (Apr.): 141–51.

—— (1983*a*). 'Cosmology, Probability and the Need to Explain Life', in N. Rescher (ed.), *Scientific Explanation and Understanding* (Lanham, Md., and London: Center for Philosophy of Science and University Press of America), 53–82.

—— (1983*b*). 'Why Not Let Life Become Extinct?', *Philosophy* (July): 329–38.

—— (1983*c*). 'Observership in Cosmology: The Anthropic Principle', *Mind* (July): 573–9.

—— (1983*d*). Review of J. King-Farlow, *Self-Knowledge and Social Relations* (New York: Science History Publications, 1978), in *Social Indicators Research* (Nov.): 425–7.

—— (1985*a*). 'Modern Cosmology and the Creation of Life', in E. McMullin (ed.), *Evolution and Creation* (Notre Dame, Ind.: University of Notre Dame Press), 91–120.

—— (1985*b*). Review of N. Rescher, *The Riddle of Existence* (Lanham, Md.: University Press of America, 1984), in *Philosophy of Science* (Sept.): 485–6.

—— (1986*a*). 'The Scientific Weight of Anthropic and Teleological Principles', in N. Rescher (ed.), *Current Issues in Teleology* (Lanham, Md., and London: Center for Philosophy of Science and University Press of America), 111–19.

—— (1986*b*). 'Anthropic Explanations in Cosmology', in A. Fine and P. Machamer (eds.), *PSA 1986*, i (Proceedings of the 1986 Biennial Meeting of the Philosophy of Science Association; Ann Arbor: Edwards Brothers), 87–95.

—— (1986c). 'Mackie on Neoplatonism's "Replacement for God" '. *Religious Studies* (Sept.): 325–42.

—— (1987). 'Probabilistic Phase Transitions and the Anthropic Principle', in J. Demaret (ed.), *Origin and Early History of the Universe* (Liège: Presses de L'Université de Liège), 439–44.

—— (1988a). 'No Inverse Gambler's Fallacy in Cosmology', *Mind* (Apr.): 269–72.

—— (1988b). 'The Prerequisites of Life in Our Universe', in G. V. Coyne, M. Heller, and J. Zycinski (eds.), *Newton and the New Direction in Science* (Vatican City: Vatican Observatory), 229–58. Repr. in Gale and Pruss (2002).

—— (1988c). 'How to Draw Conclusions from a Fine-Tuned Universe', in R. J. Russell, W. R. Stoeger, and G. V. Coyne (eds.), *Physics, Philosophy and Theology* (Vatican City: Vatican Observatory), 297–311 (Proceedings of the papal Newton Tercentenary workshop at Castel Gandolfo; distributed by University of Notre Dame Press).

—— (1989a). *Universes* (London and New York: Routledge; paperback edn., slightly revised, 1996).

—— (1989b). 'The Leibnizian Richness of our Universe', in N. Rescher (ed.), *Leibnizian Inquiries* (Lanham, Md., and London: Center for Philosophy of Science and University Press of America), 139–46.

—— (1989c). 'The Need to Generate Happy People', *Philosophia* (May): 29–33.

—— (1989d). 'Demons, Vats and the Cosmos', *Philosophical Papers* (Sept.): 169–88.

—— (1989e). 'Risking the World's End'. *Bulletin of the Canadian Nuclear Society* (May): 10–15. Repr. in *Interchange* (Spring 1990): 49–58.

—— (ed.) (1990a). *Physical Cosmology and Philosophy* (New York: Macmillan; expanded edn. from Prometheus Books, Amherst, NY, 1998, as *Modern Cosmology and Philosophy*).

—— (1990b). 'Is the End of the World Nigh?', *Philosophical Quarterly* (Jan.): 65–72.

—— (1990c). Critical Notice of J. Faye, *The Reality of the Future* (Odense: Odense University Press, 1989), in *Dialogue*, 29/3: 441–5.

—— (1991a). 'Ensuring Two Bird Deaths with One Throw', *Mind* (Jan.): 73–86.

—— (1991b). Review of A. C. Lloyd, *The Anatomy of Neoplatonism* (Oxford: Clarendon Press, 1990), in *Philosophical Books* (Apr.): 78–80.

222

—— (1992*a*). 'The Doomsday Argument', *The Mathematical Intelligencer* (Spring): 48–51.

—— (1992*b*). 'Doomsday Revisited', *Philosophical Quarterly* (Jan.): 85–9.

—— (1992*c*). 'Time and the Anthropic Principle', *Mind* (July): 521–40.

—— (1992*d*). 'Design and the Anthropic Principle', *Biology and Philosophy* (July): 349–54.

—— (1993*a*). Review of S. Weinberg, *Dreams of a Final Theory* (London: Hutchinson, 1993) and P. W. Atkins, *Creation Revisited* (Oxford: Freeman, 1992), in *The Times Literary Supplement* (29 Jan.): 3–4.

—— (1993*b*). 'Doom and Probabilities', *Mind* (July): 489–91.

—— (1993*c*). 'A Spinozistic Vision of God', *Religious Studies* (Sept.): 277–86.

—— (1993*d*). 'Creation Stories, Religious and Atheistic', *International Journal for Philosophy of Religion* (Oct.): 65–77. Repr. in R. Varghese and C. Matthews (eds.), *Cosmic Beginnings and Human Ends* (Chicago: Open Court, 1995), 337–51.

—— (1994*a*). Review of D. J. O'Meara, *Plotinus: An Introduction to the Enneads* (Oxford: Clarendon Press, 1993), in *Philosophical Books* (Apr.): 102–3.

—— (1994*b*). Review of D. Bohm and B. J. Hiley, *The Undivided Universe* (London: Routledge, 1993) and of S. W. Hawking, *Black Holes and Baby Universes* (London: Bantam, 1993), in *London Review of Books* (12 May): 15–16.

—— (1994*c*). 'Testing the Doomsday Argument', *Journal of Applied Philosophy*, 11/1: 31–44.

—— (1994*d*). 'Anthropic Prediction', *Philosophia* (July): 117–44.

—— (1994*e*). 'Cosmology: A Philosophical Survey', *Philosophia* (Dec.): 3–27.

—— (1995*a*). 'Cosmology', 'Cosmos', 'Finite/Infinite', 'World, 'Why There is Something', in J. Kim and E. Sosa (eds.), *A Companion to Metaphysics* (Oxford: Blackwell).

—— (1995*b*). Review of F. J. Tipler, *The Physics of Immortality* (London: Macmillan, 1995), in *London Review of Books* (23 Mar.): 7–8.

—— (1996*a*). *The End of the World: The Science and Ethics of Human Extinction* (London and New York: Routledge; paperback edn., slightly revised and with new preface, 1998).

—— (1996*b*). Review of W. L. Craig and Q. Smith, *Theism, Atheism and Big Bang Cosmology* (Oxford: Oxford University Press, 1993), in *Zygon* (June): 345–9.

—— (1996*c*). Review of S. W. Hawking and R. Penrose, *The Nature of Space and Time* (Princeton: Princeton University Press, 1996), in *London Review of Books* (1 Aug.): 18–19.

Bibliography

—— (1996*d*). 'A Difficulty for Everett's Many-Worlds Theory', *International Studies in the Philosophy of Science* (Oct.): 239–46.

—— (1997*a*). 'The End of the World is Not Nigh', *Nature* (22 May): 338–9.

—— (1997*b*). Review of F. J. Dyson, *Imagined Worlds* (Boston: Harvard University Press, 1997), in *London Review of Books* (5 June): 10–11.

—— (1997*c*). 'A Neoplatonist's Pantheism', *The Monist* (Apr.): 218–31.

—— (1997*d*). 'Observer-Relative Chances and the Doomsday Argument', *Inquiry* (Dec.): 427–36.

—— (1997*e*). 'The Anthropic Principle Today', in R. Hassing (ed.), *Final Causality in Nature* (Washington DC: Catholic University of America Press), 163–87.

—— (1998*a*). Review of M. Rees, *Before the Beginning: Our Universe and Others* (London: Simon & Schuster, 1998) and of L. Smolin, *The Life of the Cosmos* (London: Weidenfeld & Nicolson, 1997), in *London Review of Books* (1 Jan.): 26–8.

—— (1998*b*). Review of D. Garrett (ed.), *The Cambridge Companion to Spinoza* (Cambridge: Cambridge University Press, 1996), in *Philosophical Books* (July): 163–5.

—— (1998*c*). 'Cosmology and Theology', in the all-electronic *Stanford Enyclopedia of Philosophy*, at http://plato.stanford.edu/info.html

—— (1998*d*). Review of P. Edwards, *Reincarnation: a Critical Examination* (Amherst, N.Y.: Prometheus Books, 1996), in *Philosophical Books* (Oct.): 275–8.

—— (ed.) (1998*e*). *Modern Cosmology and Philosophy* (Amherst, NY: Prometheus Books; expanded edn. of (1990*a*).

—— (1999). 'Risking Human Extinction', in D. M. Hayne (ed.), *Human Survivability in the Twenty-First Century* (Toronto: University of Toronto Press), 117–29 (series 6, vol. 9, of *Transactions of the Royal Society of Canada*).

—— (2000*a*). 'Our Place in the Cosmos', *Philosophy* (Jan.): 5–24.

—— (2000*b*). 'The Divine Mind', in A. O'Hear (ed.), *Philosophy, the Good, the True and the Beautiful* (Cambridge: Cambridge University Press), 73–89.

—— (2000*c*). 'Intelligent Life in the Universe', in S. J. Dick (ed.), *Many Worlds* (Philadelphia and London: Templeton Foundation Press), 119–32.

—— (2001*a*). 'Anti Anti-Realism', in W. Sweet (ed.), *Idealism, Metaphysics and Community*). London: Ashgate), 111–17.

—— (2001*b*). 'The Meaning of Design'. Forthcoming in J. B. Miller (ed.), *Cosmic Questions,* published by the New York Academy of Sciences. To be reprinted in N. Manson (ed.), *God and Design* (London: Routledge, 2002).

224

Levine, M. (1994). *Pantheism* (London and New York: Routledge).

Lewis, D. (1983). *Philosophical Papers*, i (New York and Oxford: Oxford University Press).

—— (1986). *On the Plurality of Worlds* (Oxford: Basil Blackwell).

Linde, A. D. (1990). *Inflation and Quantum Cosmology* (San Diego, Calif.: Academic Press).

Locke, John (1700). *An Essay Concerning Human Understanding* (4th edn.; repr. 1982, ed. P. H. Nidditch, Oxford: Clarendon Press).

Lockwood, M. (1989). *Mind, Brain and the Quantum* (Oxford: Blackwell).

—— (1993). 'Dennett's Mind', *Inquiry* (Mar.): 59–72.

Loptson, P. (1988). 'Spinozist Monism', *Philosophia* (Apr.): 19–38.

Mackie, J. L. (1976). *Problems from Locke* (Oxford: Clarendon Press).

—— (1977). *Ethics: Inventing Right and Wrong* (Harmondsworth: Penguin Books).

—— (1982). *The Miracle of Theism* (Oxford: Clarendon Press).

McTaggart, J. M. E. (1901). *Studies in Hegelian Cosmology* (Cambridge: Cambridge University Press).

—— (1927). *The Nature of Existence*, ed. C. D. Broad (Cambridge: Cambridge University Press).

Marshall, I. N. (1960). 'ESP and Memory: A Physical Theory', *British Journal for the Philosophy of Science* (Feb.): 265–86.

—— (1989). 'Consciousness and Bose–Einstein Condensates', *New Ideas in Psychology*, 7/1: 73–83.

Moore, A. W. (1990). *The Infinite* (London and New York: Routledge).

—— (1995). 'A Brief History of Infinity', *Scientific American* (Apr.): 112–16.

Moravec, H. P. (1988). *Mind Children: The Future of Robot and Human Intelligence* (Cambridge, Mass.: Harvard University Press).

—— (1989). 'Human Culture: A Genetic Takeover Underway', in C. G. Langton (ed.), *Artificial Life* (Redwood City, Calif.: Addison-Wesley), 167–99.

—— (1999). *Robot: Mere Machine to Transcendent Mind* (New York and Oxford: Oxford University Press).

Nagel, T. (1979). *Mortal Questions* (Cambridge: Cambridge University Press).

Nozick, R. (1981). *Philosophical Explanations* (Oxford: Clarendon Press).

—— (1989). *The Examined Life* (New York: Simon & Schuster).

Parfit, D. (1984). *Reasons and Persons* (Oxford: Oxford University Press).

—— (1992). 'The Puzzle of Reality', *The Times Literary Supplement* (3 July): 3–5.

Bibliography

Parfit, D. (1998). 'Why Anything? Why This?', a two-part article in *London Review of Books* (27 Jan.): 24–7, and (5 Feb.): 22–5.

Penrose, R. (1987). 'Minds, Machines and Mathematics', in C. Blakemore and S. Greenfield (eds.), *Mindwaves* (Oxford: Blackwell), 259–76.

—— (1989). *The Emperor's New Mind* (Oxford: Oxford University Press).

—— (1994). *Shadows of the Mind* (Oxford: Oxford University Press).

—— with Shimony, A., Cartwright, N., Hawking, S., and Longair, N. (1997). *The Large, the Small and the Human Mind* (Cambridge: Cambridge University Press).

Plantinga, A. (1980). *Does God have a Nature?* (Milwaukee, Wis.: Marquette University Press).

—— and Grim, P. (1993). 'Truth, Omniscience and Cantorian Arguments: An Exchange', *Philosophical Studies* (Sept.): 267–306.

Polkinghorne, J. (1986). *One World: The Interaction of Science and Theology* (Princeton: Princeton University Press).

—— (1994). *The Faith of a Physicist* (Princeton: Princeton University Press).

Pritchard, D. B. (1994). *The Encyclopedia of Chess Variants* (Godalming: Games and Puzzles Publications).

—— (2000). *Popular Chess Variants* (London: Batsford).

Putnam, H. (1997). 'Thoughts Addressed to an Analytical Thomist', *The Monist* (Oct.): 487–99.

Redhead, M. (1995). *From Physics to Metaphysics* (Cambridge: Cambridge University Press).

Rees, M. J. (1997). *Before the Beginning: Our Universe and Others* (Reading, Mass.: Addison-Wesley).

—— (1999). *Just Six Numbers: The Deep Forces that Shape the Universe* (London: Weidenfeld & Nicolson).

Rescher, N. (1984). *The Riddle of Existence* (Lanham, Md., and London: University Press of America).

—— (2000). *Nature and Understanding* (Oxford: Clarendon Press).

Rice, H. (2000). *God and Goodness* (Oxford: Oxford University Press).

Robinson, J. A. T. (1963). *Honest to God* (London: SCM Press).

Rozental, I. L. (1980). 'Physical Laws and the Numerical Values of Fundamental Constants'. *Soviet Physics: Uspekhi* (June): 293–305.

—— (1988). *Big Bang, Big Bounce* (Berlin: Springer-Verlag).

Rucker, R. (1983). *Infinity and the Mind* (New York: Bantam).

Russell, B. (1903). *Principles of Mathematics* (Cambridge: Cambridge University Press).

—— (1914). *Our Knowledge of the External World* (Chicago: Open Court).

—— (1927). *The Analysis of Matter* (London: Kegan Paul).

Sartre, Jean-Paul (1957). *The Transcendance of the Ego*, tr. F. Williams and R. Kirkpatrick (New York: Noonday Press).

Schrödinger, E. (1958). *Mind and Matter* (Cambridge: Cambridge University Press).

—— (1964). *My View of the World* (Cambridge: Cambridge University Press).

Seager, W. (1999). *Theories of Consciousness* (London and New York: Routledge).

Searle, J. R. (1984). *Minds, Brains and Science*, Reith Lectures published in *The Listener*.

—— (1997). *The Mystery of Consciousness* (London: Granta Books).

Sheldrake, R. (1987). *A New Science of Life* (London: Paladin).

Shimony, A. (1988). 'The Reality of the Quantum World', *Scientific American* (Jan.): 46–53.

—— (1997). 'On Mentality, Quantum Mechanics and the Actualization of Potentialities', in Penrose *et al.* (1997: 144–69).

Shirley, S. (ed. and tr.) (1982). *The Ethics of Spinoza, and Selected Letters* (Indianapolis: Hackett).

Shoemaker, S. (1969). 'Time without Change', *Journal of Philosophy* (19 June): 363–81.

Simoni, H. (1997*a*). 'Omniscience and the Problem of Radical Particularity', *International Journal for Philosophy of Religion* (Aug.): 1–22.

—— (1997*b*). 'Divine Passibility and the Problem of Radical Particularity: Does God Feel your Pain?', *Religious Studies* (Sept.): 327–47.

Smart, J. J. C. (1967). 'Time', in P. Edwards (ed.), *The Encyclopedia of Philosophy*, vii (New York: Macmillan and Free Press), 126–34.

—— (1989). *Our Place in the Universe* (Oxford: Blackwell).

Smolin, L. (1997). *The Life of the Cosmos* (London: Weidenfeld & Nicolson).

Spiller, T., and Clark, T. (1986). 'SQUIDs: Macroscopic Quantum Objects', *New Scientist* (4 Dec.): 36–40.

Sprigge, T. L. S. (1971). 'Final causes, I', *Proceedings of the Aristotelian Society*, supplementary vol. 45: 149–70.

—— (1983). *The Vindication of Absolute Idealism* (Edinburgh: Edinburgh University Press).

—— (1984). *Theories of Existence* (Harmondsworth: Penguin Books).

—— (1997). 'Pantheism', *The Monist* (Apr.): 191–217.

Stoerig, P., and Cowey, A. (1989). 'Wavelength Sensitivity in Blindsight', *Nature* (21 Dec.): 916–18.

Strawson, G. (1994). *Mental Reality* (Cambridge, Mass.: The MIT Press).

Swinburne, R. (1970). *The Existence of God* (Oxford: Clarendon Press).

Bibliography

Swinburne, R. (1977). *The Coherence of Theism* (Oxford: Clarendon Press).

Tillich, P. (1953–63). *Systematic Theology* (London: Nisbet).

Tipler, F. (1994). *The Physics of Immortality* (New York: Doubleday).

Tryon, E. P. (1973). 'Is the Universe a Vacuum Fluctuation?', *Nature* (14 Dec.), 396–7.

Unger, P. (1984). 'Minimizing Arbitrariness: Toward a Metaphysics of Infinitely Many Isolated Concrete Worlds', *Midwest Studies in Philosophy*, ix (Minneapolis: University of Minnesota Press), 29–51.

Vesey, G. N. A. (ed.) (1964). *Body and Mind* (London: George Allen & Unwin).

Vilenkin, A. (1982). 'Creation of Universes from Nothing', *Physics Letters B* (4 Nov.), 25–8.

Ward, K. (1982). *Rational Theology and the Creativity of God* (Oxford: Blackwell).

—— (1996a). *Religion and Creation* (Oxford: Clarendon Press).

—— (1996b). *God, Chance and Necessity* (Oxford: Oneworld Publications).

Whitehead, A. N. (1927). *Religion in the Making* (Cambridge: Cambridge University Press).

—— (1938). *Modes of Thought* (Cambridge: Cambridge University Press).

—— (1978). *Process and Reality* (corrected edn.; New York: Macmillan).

Wigner, E. P. (1960). 'The Unreasonable Effectiveness of Mathematics in the Natural Sciences', *Communications in Pure and Applied Mathematics*, 13/1: 1–14.

Williams, D. C. (1951). 'The Myth of Passage', *Journal of Philosophy*: repr. in Gale (1967: 98–116).

Wittgenstein, L. J. J. (1976). *Lectures on the Foundations of Mathematics*, ed. C. Diamond (Brighton: Harvester).

—— (1978). *Remarks on the Foundations of Mathematics*, ed. G. H. von Wright, R. Rhees, and G. E. M. Anscombe (Oxford: Blackwell).

Wolf, A. (ed. and tr.) (1910). *Spinoza's Short Treatise on God, Man and his Well-Being, and a Life of Spinoza* (London: Adam & Charles Black).

Wynn, M. (1999). *God and Goodness* (London and New York: Routledge).

Zohar, D. (1996). 'Consciousness and Bose-Einstein Condensates', in S. R. Hameroff, A. W. Kaszniak, and A. C. Scott (eds.), *Towards a Science of Consciousness* (Cambridge, Mass.: The MIT Press), 439–50.

Index of Names

Index

Index of Subjects

Index